180 Weeks in the War on Terror

Steven Stalinksy

ميمري
MEMRI
BOOKS

Washington, D.C.

180 WEEKS IN THE WAR ON TERROR

Published in the United States of America
by MEMRI Books

Paperback ISBN: 978-1-7344283-2-2
E-book ISBN:978-1-7344283-3-9
Library of Congress Control Number:2021909471

Contents

Introduction:
180 Weeks in the War on Terror

From 2003 to 2008, Steven Stalinsky penned a weekly column for *The New York Sun*, covering a wide range of aspects of the post-9/11 world, the war in Iraq, and the War on Terror.

His columns capture an important era that began shortly after the launch of the War on Terror and continued for five years. As the 20th anniversary of 9/11 approaches, it is worth taking another look at these weekly columns; they mark historic milestones and remind us of events and developments that may already be forgotten – including the beginning of the ongoing struggle against radical Islamism.

The topics he wrote on ranged from common themes in Arab and Iranian media on the War on Terror and early jihadi use of the Internet to translated statements by influential Arab and Muslim figures, and much more original analysis and translations.

Most importantly, he wrote about numerous issues concerning extremism and terrorism years before they became as widely understood and widely discussed as they are today.

For example, he explained in detail how terrorists were beginning to use social media platforms, such as YouTube, for uploading and disseminating propaganda videos – among them graphic videos of beheadings and executions, for which he coined the phrase "jihadi porn," a phrase quoted by others to this day.

He wrote about conspiracy theories in the Middle East; on Islamic extremism on satellite TV programs broadcast in the West, including those of Al Jazeera, as an early critic raising the alarm and spotlighting its Muslim Brotherhood connection; on the rise of Islamism in Turkey, propelled by then-Turkish Prime Minister Recep Tayyip Erdogan and his Justice and Development (AKP) party; and more.

His columns also included several series devoted to such topics as Hezbollah's Al-Manar TV – a Specially Designated Global Terrorist Entity – and Hamas's Al-Aqsa TV, which were widely used at briefings on Capitol Hill during this time. These and other of his columns have been cited in official U.S., Canadian, U.K., and U.N. documents and reports.

Over the weeks and years, some of his columns had an international impact – from the very highest echelons. The Saudi royal family was incensed by his April 21, 2004 column "The Saudi Royal Family and the Jews" about blatant Saudi anti-Semitism, and Saudi Princess Fahda bint Saud responded in the Saudi media. She wrote, inter alia, that "there is no denying that the September 11 attacks have served the interests of the Zionists more than they have any other group's" and "If the pro-Israeli lobby did not influence the U.S. media, then the U.S. could hardly prefer the interests and friendship of a small country like Israel to that of 1.5 billion Muslims." Of Mr. Stalinsky himself, she said that he "must have been in great hurry to publish" the column, "as his references are a collection of pieces of articles taken out of their contexts."

Another was the response to his April 7, 2004 column titled "Egypt's Al-Azhar's Jihad Against the U.S. War On Terrorism" criticizing the extremism at Al-Azhar, the most prestigious center Islamic center of learning in the Sunni Muslim world. None other than the institution's head, Sheikh Dr. Muhammad Sayyid Tantawi, whose authority for Sunni Muslims is like that of the pope for Roman Catholics, struck back, writing in the Arab media a week later that Stalinsky "mix[es] up the cards, distort[s] the truth, and smear[s] the reputation of honorable people" and that his suggestions for reform and oversight of the institution he heads "is in fact an obscenity that no one should pay attention to."

Jihad Al-Khazen, a writer with a great deal of influence in the Arab world with alleged connections with bin Laden and Al-Qaeda, wrote a column mocking John Bolton, at the time Under Secretary of State for Arms Control and International Security Affairs, for looking "inappropriate... like a walrus." This was mentioned by Stalinsky in his June 22, 2005 column "Arab, Iranian Press Attack Bolton" – and Al-Khazen, whose column was at that time among the most widely read across the Middle East, felt compelled to retaliate in a July 2005 column with puerile insults: "Stalinsky['s] name caught my eye, as it is derived from Stalin; it reminded me of Stalinist propaganda and... disinformation."

Following the publication of his July 20, 2005 column about Sheikh Abd Al-Rahman Al-Sudayyis, the government-appointed imam of the Grand Mosque in Mecca was barred from entering Canada. Mr. Stalinsky quoted Sudayyis as stating in a sermon: "Oh Allah, liberate our Al-Aqsa Mosque from the defilement of the occupying and brutal Zionists ... Oh Allah, punish the occupying Zionists and their supporters from among the corrupt infidels. Oh Allah, scatter and disperse them, and make an example of them for those who take heed." Previous sermons, he explained, were characterized by animosity towards non-Muslims, calling Jews "scum of the earth" and "monkeys and pigs" who should be "annihilated," and insistence that "worshippers of the cross" and "idol-worshipping Hindus" must be fought.

In a May 2004 column, he quoted Sheikh Taj Al-Din Hamed Abdallah Al-Hilali, mufti of Australia and New Zealand who was considered a liberal, as saying in a sermon: "Don't be surprised if one day you hear the muezzin calling for praying and saying 'Allah Akbar' from the top of the White House. September 11 is Allah's work against oppressors." Mr. Stalinsky's exposure of these and other statements supportive of jihad, martyrdom operations, and terrorist organizations with which he met, including Hezbollah, led Australian political leaders to widely condemn him. He was later forced to step down from his post.

Mr. Stalinsky also wrote about Adam Gadahn, aka Azzam the American, the California boy who became a media trailblazer for Al-Qaeda, and its American face, and brought jihad into the information era. This formed the basis of the book he would publish a decade later – *AMERICAN TRAITOR: The Rise and Fall of Al-Qaeda's U.S.-Born Leader Adam Gadahn.* His September 13, 2006 column on Gadahn, titled A Jewish Musician's Son Joins Al Qaeda's Ranks, was cited the following week in a Congressional hearing on "The Homeland Security Implications Of Radicalization."

At the time these columns were written, between 2003 and 2008, they were used and cited by prominent figures and media in the U.S. and internationally. Today, this collection – comprising 180 weeks of the War on Terror – makes an important contribution to the body of literature relating to the War on Terror.

The following are random thoughts on issues concerning the War on Terror, written every week for the New York Sun between March 2004 and May 2007.

2004

Sympathy for the Terrorist

March 24, 2004

"The day in which I die as a shahid will be the happiest day of my life," Sheikh Ahmed Yassin told the Palestinian daily Al-Quds in 1998. The interview was given following Yassin's tour of Saudi Arabia and Iran, in which a jubilant Yassin boasted of his meeting with the king of Saudi Arabia, who "supports the path of jihad." The Saudi Press Agency reported that upon hearing of Yassin's death, Saudi King Fahd sent his "warmest condolences" to Yassin's family. In the 1998 interview, Yassin explained that Saudi Arabia and Iran "promised financial assistance...to support the Palestinian people...and the families of martyrs..."

Iran's response to Yassin's killing included a statement by the ruling cleric, Ayatollah Ali Khamenei: "The blood of Ahmed Yassin will nourish the Islamic resistance...The Zionist regime is doomed to annihilation." Iran's Expediency Council head, Akbar Hashemi Rafsanjani, said no death could be more fitting for Sheikh Yassin, and predicted that the "dastardly" killing will expedite the Palestinian "ultimate victory" over Israel.The Iranian president, Mohammed Khatemi, added, "I expect the international community to condemn the criminal Zionist regime and their evil supporters, particularly the U.S."

Other leaders condemning the killing include the Egyptian president, Hosni Mubarak, who, according to Egypt's State Information Service, expressed, "Deep sorrow for the death of Sheikh Yassin," and "warned of the dire consequences of the completely meaningless and miscalculated Israeli action."

According to the Yemen News Agency, Yemen's parliament speaker, Abdullah Al-Ahmar, sent his condolences to the chairman of the legislative council of Palestine, Roohi Futooh, and the head of the political office of Hamas, Khaled Misha'al. Mr. al-Ahmar said this assassination will fuel Palestinian resistance to "wipe out the Zionist terror and make the world safe from it."

Leading Muslim religious figures called for Israel to be "punished" with "jihad." The sheikh of Al-Azhar University, Muhammad Sayyed Tantawi, called on Arabs and Muslims "to punish those who perpetrated this heinous crime," while prominent Islamist leader Sheikh Yousef Al-Qaradhawi urged Muslims to rally behind the

Palestinian resistance, according to IslamOnline.com: "Only force and Jihad work with those aggressors...He met his Creator satisfied and pleased [with his decades of resistance and struggle]. Truly, dying a martyr is the best death a Muslim can get."

The Palestinian Arab reaction also condemned the attack. Sheikh Hamid Al-Bitawi, the imam of Al-Aqsa mosque and head of the Palestinian Scholars League, also speaking to IslamOnline, said that it is now "a religious duty to liberate Palestine." Yasser Arafat, according to the Palestinian National Authority's press information center, was quoted as saying, "The Palestinians will not waver in their legitimate and national rights, and would continue their resistance against the Israeli occupation."

According to the Web site www.palestine-info.net, Mr. Arafat said Yassin will be remembered as "The Fighting Hero Sheikh."

Hamas's military wing, the Izz Al-Din Al-Qassam Brigades, issued a threat to America and Israel: "Whoever ordered the assassination of Ahmed Yassin ordered thereby the murder of hundreds of Zionists...The Zionists would have never undertaken this operation without the approval of the terrorist American administration that must bear the responsibility for this crime...Not only will all fighting Palestinian factions participate in the response, but Muslims from the entire Muslim world will have the honor of participating in it."

Sheikh Yassin had also threatened America in February 2003, during the buildup to actions to depose Saddam Hussein. The Hamas Web site posted a communiqué in which Yassin called on the Islamic nation "to strike at Western interests everywhere...The hatedrenched West, headed by America, declared today a crusade on the [Islamic] nation and on the Muslim belief when it gathered to strike Iraq."Yassin demanded that Iraq "open its borders to all Muslims across the world so that they can play their part in the defensive battle of the [Islamic] nation." Iraqis were urged byYassin to prepare for martyrdom, and he stressed that "if Iraq is conquered, jihad will be the personal obligation of every Muslim man and woman, and there will be no alternative to Muslims threatening the interests of the hostile Americans and Westerners and striking at them everywhere."

In America, the Council on American-Islamic Relations, or CAIR, issued a press release which condemned the assassination of Yassin. "We condemn this violation of international law as an act of state terrorism by Ariel Sharon's out-of-control government...The international community must now take concrete steps to help protect the Palestinian people against such wanton Israeli violence..."

The Syrian Blood Libel – Passover 2004

March 31, 2004

The Syrian-produced Hezbollah TV series "Al-Shatat," which ran earlier this year during Ramadan, represents a long line of Syrian Blood Libels. In the 20th episode of "Al-Shatat," a "rabbi" says, "We want the blood of a Christian child before Passover, for the matzos." A neighborhood boy named Nathan, who looks to be about 8 years old, is brought into the room, his throat is slit as he cries "mama, mama" and his blood is poured into a metal basin. The next scene has the rabbi wishing a "Jewish man" a good Passover and offering him matzo — "tastier and holier because it was kneaded with pure blood."

Twenty-one years earlier, the Syrian minister of defense, Mustafa Tlass, who is considered one of the pillars of the Syrian Baathist regime, released the first edition of his book, "The Matzah of Zion." The 1983 book retold the notorious Damascus Blood Libel of 1840 in which the Jewish community in Damascus was accused of murdering a Christian priest named Toma and his servant to use their blood in making matzahs.

The second edition of Mr. Tlass's book was published in 1986, with added footnotes and pictures, including one of the tombstone of Father Toma's grave with the inscription that he was killed by Jews. Mr.Tlass's book has become influential in international anti-Semitic circles as a reliable source of information about Jewish ritual murder and has been published in multiple languages, including French and Italian.

By 2002, Mr. Tlass's book had been reprinted eight times, and was one of the most popular books at the Syrian Book Fair that year. According to its publisher, the reason for its popularity is "the will of the next generation to know about the Jews, how they harm Arabs and others." Mr.Tlass explained that in each reprint he will add new documents "which will shed light on the distortion of the Torah [by the Jews] and on criminal Jewish religious rituals." Similarly, in the 1986 edition's introduction, citing the Koran, Mr. Tlass wrote, "I intend to illuminate some of the secrets of the Jewish religion by describing the actions of the Jews...and the implementation of the Talmudic precepts compiled in the Diaspora by their rabbis who distorted the principles of the Jewish belief."

In 2001, it was reported in the Egyptian press that Mr. Tlass's book will be made into a movie in which he will receive 5% of its profits. The movie's producer will be the Egyptian Munir Radhi and the Egyptian cast will reportedly be led by Omar Sharif. According to Mr. Radhi, the primary goal of the movie is "to respond to all the Zionist films distracted by the American film industry," notably "Schindler's List." During a 2003 interview, Mr. Tlass explained that his book is not anti-Semitic and compared such charges to those against French Holocaust-denier Roger Garaudy, because "world Zionism opposes anyone who speaks the truth."

Following the airing of "Al-Shatat" this year, there has been a growing movement in Europe to block Hezbollah TV. In late 2003, Australia was the first Western nation to announce the suspension of the channel Al-Manar from its airwaves following a government viewing of "Al-Shatat" and with a subsequent investigation of the channel alleging criminal activities of broadcasting programs that "perpetuate hatred against or gratuitously vilify any person or group."

France announced in February 2004 that it is considering banning Al-Manar. Le Monde reported that following the French Prime Minister's viewing of "Al-Shatat," he stated that the government would take action to stop its broadcasting. Earlier, in November 2000, a French court subpoenaed the editor of the Egyptian government paper Al-Ahram for printing a Blood Libel strikingly similar to the one told by Mr. Tlass.

During a session of parliament in Sweden on March 18 in reference to "Al-Shatat," lawmaker Mikael Oscarsson of the Christian Democratic Party asked Prime Minister Persson to put an end to Al-Manar broadcasts in Sweden, which are "appalling propaganda of incitement" that "can only be compared with that of the Nazis."

While America condemned the series during its airing — as spokeswoman for the State Department's Bureau of Near Eastern Affairs, Rhonda Shore, said, "Such venomous anti-Semitism has absolutely no place in the civilized world" — there is no government effort to block Al-Manar in America. Surprisingly, during an Al-Manar broadcast on December 14, 2003, it was reported that Hezbollah TV actually has a correspondent stationed in Washington, D.C.

The Web site for Al-Manar, www.manartv.com, is also hosted in America by a company in Hoboken, N.J. To view the video of "Al-Shatat," visit http://memri.org/video/index.html.

Egypt's Al-Ahzar's Jihad Against the U.S. War on Terrorism

April 7, 2004

This week President Bush will host Egyptian President Hosni Mubarak at his Texas ranch to discuss issues such as the war on terrorism. Immediately after 9-11, a group of leading clerics and professors declared jihad against the U.S. at the most prestigious Islamic center of learning in the Muslim world - the Egyptian government supported Al-Ahzar. The issue of extremism at Al-Ahzar has emerged as one aspect of the war on terrorism which has not received much attention.

Sheikh of Al-Ahzar, one of the most coveted positions of authority in Sunni Islam, from 1929-1935 was Muhammad al-Ahmadi al-Zawahiri, grandfather of Al-Qa'ida's Ayman al-Zawahiri. More recently Dr. Muhammad Sayyed Tantawi, who previously served as Mufti of Egypt until he was appointed by the government as Sheikh of Al-Ahzar, has led the call in Egypt for jihad against U.S. forces in Iraq. At a press conference last year, Tantawi called on the Iraqi people to "continue its jihad in defense of religion... whether martyrdom operations [i.e. suicide operations] or [by] any other means." He encouraged volunteers from Muslim countries to go to Iraq "to support the jihad ... because resistance to oppression is an Islamic obligation..." During a Friday sermon at Al-Azhar the following day, Tantawi added: "The American aggression against Iraq is not acceptable to Islamic law ... The Iraqi people must defend itself... because it is a jihad that is authorized by Islamic law... The gates of jihad are open until the Day of Judgment, and he who denies this is an infidel ..."

Before returning to Egypt within a month of 9-11, the Al-Ahzar representative in the U.S. and then Imam of the New York Islamic Center, Sheikh Muhammad Al-Gamei'a, stated that the Jews were behind 9-11 - not Osama bin Laden, and that the U.S.'s attack on Afghanistan would lead to "the end of America the oppressor." He explained that following the annihilation of America, "the future of the Muslims in the U.S. will be glorious."

When the U.S. launched its attack on the Taliban, the official spokesman of the

Al-Ahzar clerics, Dr. Yahyah Isma'il declared: "This war is a war against Muslims and Islam. Everyone must offer help to the mujahideen brothers in Afghanistan." Sheikh Ali Abu Al-Hassan, head of Al-Azhar's Religious Ruling Committee, added, "If the enemy sets foot on the lands of Islam, he must be fought... Islam urges us to set out on a jihad ... this instance is an obligation for all Muslims... we are obliged to help them and to defend them as best we can from the cruelty of the American attacks. Islam has commanded us ... to fight the jihad alongside them against the polytheists..." Dr. Abd Al-Azim Al-Mut'ani, Al-Azhar University, lecturer compared U.S. actions as "similar to the first Crusader wars" while Dr. Abd Al-Hay Al-Farmawi, another lecturer, referred to the American campaign in Afghanistan as follows: "America is doing this without presenting any proof of Osama bin Laden's guilt. This proves that this war, as Bush said, is a crusader war..."

Al-Azhar's Islamic Research Complex has also called upon all Muslims to launch jihad in response to the U.S. actions in Iraq: "Jihad is an individual duty for all Muslims if the U.S. launches a war against Iraq... Our Arab and Muslim nation will face new crusades that aim to deprive us of our homeland, doctrine and dignity." Sheikh 'Ali Abu Al-Hassan, also stated it is an obligation to fight American forces in Iraq and "the blood of the American and British soldiers is permitted in this instance... the Muslims killed will be considered Shahids [martyrs]."

According to reports Al-Hassan was fired for making the statement, while another communiqué issued by the Islamic Center for Research about Iraq and approved by Tantawi added: "... it is in accordance with logic and with Islamic religious law that if the enemy raids the land of the Muslims, jihad becomes an individual's commandment, applying to every Muslim man and woman, because our Muslim nation will be subject to a new Crusader invasion ..."

The issue of extremism at Al-Ahzar is related to both the war on terrorism, as well as the broader issue of reform in the Middle East and, in particular, Egypt. While the religious establishment of Al-Ahzar has been surprisingly open to the concept of reform, Egypt's mufti explained last month – it must come from within. The need to change Al-Ahzar's religious discourse has been discussed by Egyptian reformists such as intellectual and poet Ahmed Abd Al-M'uti Higazi who explained "...Those who quote [religious scriptures] and impose the word [namely, the chief clerics] are the ones responsible for producing fundamentalist terror ..." Higazi added, "The positions of the sheikh of Al-Azhar ... were established by the state, but they serve only to root the principle of quote-hear-obey. They kill creativity and lead to the atrophy of the Arab mind."

Arab Reform: Qatar Model

April 14, 2004

While the Egyptian government and Saudi royal family, generally known as the main American allies in the Arab world, have been hostile to American reform initiatives, the tiny Gulf state of Qatar, best known for being the headquarters of al-Jazeera TV, has emerged as the most supportive Arab nation of American reform initiatives in the Middle East.

As the Egyptian scholar Osama Al-Ghazali Harb wrote this week: "Reform on the regional level requires countries of a certain region to derive support, and receive inspiration from each other. In this respect, one 'model' country can play an important role in stimulating the region as a whole." The political and religious leadership in Qatar are, in fact, emerging as a progressive force in the Arab world that can serve as that model. It is setting the standard for other nations in the region to follow dealing with democracy, reform in Islam, education, economics, and the rights of women.

In early 2004, Qatar announced it would host a series of conferences on the topic of reform in the Middle East. The first focused on educating Arabs about elections and the participation of women. Another conference dealt with human rights in educational curricula. This was followed by another conference organized by the Qatari foreign ministry that focused on issues relating to development, strengthening economic growth, and as well as peace in the region.

Following up from a March 3 meeting of Arab foreign ministers in which Qatar defended American reform initiatives, Qatar's outspoken foreign minister, Hamad Bin Jassim Aal-Thani, explained two weeks later that the Arab countries are not ready to carry out reform from within and would be better off to follow the American reform initiatives. Speaking later at the Doha Cultural Festival, he added, "development will not be achieved...without enhancing democracy, human rights, equality and giving women their rights."

Mr. Aal-Thani referred to the Bush administration's democratization initiative for the Middle East, explaining, "We support reform, development, and promotion of human rights in the region." He added, "Each country [in the Arab world] can select its own model, but the basic principles of democracy, in our view, remain the freedom to choose, equality, prevalence of law and respect for human rights."

Presenting the inaugural address at the 4th Conference on Democracy and Free Trade in Doha last week, Mr. Aal-Thami said that the Arab world has been foolish to dismiss American reform initiatives. Countries in the region, he added, "can demonstrate that their efforts are not less than those countries like Malaysia, India, and Singapore, which suffered pains of the past but rose above them, and hence developed economically and politically."

He then went on to blast the Arab world for using the Palestinian issue as an excuse not to move forward. "Honesty obliges us to stress that the wrath in our region does not spring only from the Palestinian cause but goes deeper and is due to problems of our own creation that have nothing to do with the outside world - problems we have allowed to grow unremedied and unchecked..."

Discussing the importance of bringing together those involved with reform, he explained, "We placed emphasis on opening the door for dialogue between all religions and civilizations, as well as on the role of education and the rights of women. The development of free trade and advanced technologies provide support for the development of democratic principles."

Last week's conference brought others supportive of reform, including former Pakistani information minister, Mushahad Hussein Sayed, who said: "Realization has dawned on the leadership of the region that its future lies in democracy...Democracy can't be piecemeal or selective; it means respect for public opinion, the rule of law...It also means protecting the human rights of all people, irrespective of age, sex, culture, or religion."

Another speaker, the Organization of the Islamic Conference secretary-general, Abdelouahed Belkaziz, said Islam is fully compatible with the concepts of democracy although, in some countries, the way in which the law has been written has diverged from the principles of faith.

He added, "Some reform is needed, and a lack of democracy also adversely affects economic performance.

In order to attract 'economic actors' to the region, there must be justice, accountability, transparency, and the encouragement of investment."

The Saudi Royal Family and the Jews

April 21, 2004

Leading members of the Saudi royal family have been on the forefront of espousing an extreme position of hatred toward Jews, influencing the Kingdom's educational system, press, mosques, and its foreign and domestic policy.

In its first attempt to attract tourists to the country, Saudi Arabia's tourist commission, under the control of Prince Sultan bin Abd Al-Aziz, launched its Web site in March 2004, which listed those not allowed into the kingdom: "Israeli passport holders or those whose passport has an Israeli arrival/departure stamp; those who do not abide by the Saudi traditions concerning appearance and behavior; those under the influence [of alcohol]; and Jewish people."

The Saudi embassy's Washington spokesman, Nail Al-Jubeir, said that he was "stunned" when he saw the Web site, while Saudi ambassador to America, Prince Bandar bin Sultan explained that he was "embarrassed." According to a press release by the Saudi embassy, "The information on the website was not correct and as a consequence the erroneous material was removed."

The ambassador's father, Prince Sultan, who serves as secretary general of the Tourism Commission, said in a statement that the controversy was "blown out of all proportions" by American press seeking to portray the Kingdom as anti-Semitic. He added: "... It is all part of a smear campaign meant to tarnish Saudi Arabia's image."

Prince Sultan, who is also the second deputy prime minister, defense and aviation minister, and inspector general of Saudi Arabia, has been making statements against Jews for years. Following a ceremony at the Saudi Public Institution for Military Industries in June 2002, when asked about American criticism of Saudi Arabia, Prince Sultan replied to the Saudi daily Al-Sharq Al-Awsat, "It is enough to see a number of congressmen wearing Jewish yarmulkes to explain the allegations against us." According to Al-Hayat, which also reported the incident, the minister cut off a journalist, saying: "Say 'Jewish congressmen.' It is enough that in the pictures you see some congressmen wearing Jewish yarmulkes. That is sufficient." More recently, the Saudi royal family Web site, 'Ain-Al-Yaqeen, quoted Prince Sultan discussing American reform initiatives in its March 26 to April 3 issue, explaining that the American press, which is "under the Jewish influence" is using the initiative to widen the gap between Arab countries and America.

Prince Sultan's brother, the Saudi minister of interior, Prince Naif bin Abd Al-Aziz, has also made accusations against the Jews. In what has since become an infamous interview reported in 'Ain-Al-Yaqeen a year after the September 11 attacks, Prince Naif explained that Arabs were not involved in the attacks, saying, "We put big question marks and ask who committed the events of September 11 and who benefited from them. Who benefited from events of 9/11? I think they [the Jews] are behind these events."

Saudi kings have also been known to espouse extreme anti-Semitic opinions. Saudi Princess Fahda bin Saud ibn Abd Al-Aziz, who has been described as "the daughter of King Saud and the historian of her father's reign," writes occasionally in the Saudi press on conspiracy theories against the Jews and America. She has written that her father's views on the Jews and the Jewish state still serve as inspiration for the Arab and Muslim world.In one article she authored, she explained that King Saud called the Jewish state a deadly disease which would never be accepted by Arabs: "...King Saud made the right diagnosis: 'The Zionist threat is like cancer — in dealing with it neither medicine nor surgery will do any good.' This royal statement was meant to emphasize that the Arabs do not, and will not, accept an Israeli state amidst them." The article added that under the leadership of King Saud, the Saudi representative to the United Nations called for the U.N. to establish an agency "to help resettle Jews [now in Israel] in their former European homes."

The late King Faisal was also notorious for making anti-Semitic statements. He told the Egyptian magazine al-Musawwar in 1972, "While I was in Paris on a visit, the police discovered five murdered children. Their blood had been drained, and it turned out that some Jews had murdered them in order to take their blood and mix it with the bread they eat on that day."

The Saudi royal family's hatred of the Jews is now influencing its next generation. Saudi Prince Amr Muhammad Al-Faisal writes often in the Saudi press articles warning Jews in America that eventually Americans will turn against them. In one article he declared: "Dear cousins, if you hear a snap in two or three years, it will probably be the sound of the trap shutting on your collective necks. You have been warned."

Given that the Saudi royal family controls its country's press, mosques, and textbooks, they are clearly responsible for the hatred of Jews which has become closely associated with the Kingdom.

The Mufti of Australia's Jihad

April 28, 2004

'Don't be surprised if one day you hear the Muezzin calling for praying and saying 'Allah Akbar' from the top of the White House. September 11 is Allah's work against oppressors," said the mufti of Australia and New Zealand, Sheikh Taj Al-Din Hamed Abdallah Al-Hilali.

The mufti, who had enjoyed a reputation of being a liberal, had been exposed when a series of his interviews and sermons over the past two years were translated from Arabic into English in a February report by MEMRI. His statements were supportive of jihad and martyrdom operations as well as for terrorist organizations with whom he met, such as Hezbollah. Mr. al-Hilali also explained that the "axis of evil" is really President Bush, the Australian prime minister, John Howard and the British prime minister, Tony Blair.

Following the translation of Mr. al-Hilali's statements, the Australian government was highly critical of the country's leading Muslim religious authority. Mr. Howard condemned his remarks, saying, "Incitement to a jihad" is "utterly unacceptable." In another interview, Mr. Howard said, "What he has said deserves to be condemned in the strongest possible terms."

Canada's National Post newspaper reported on March 6 that Mr. al-Hilali's spokesman and other prominent Australian Muslims, including the leader of the Australian Federation of Islamic Councils, Amir Ali, challenged the accuracy of translations of the mufti's remarks, and claimed that this was a further case of Western mistreatment of Muslims. In response, the Australian foreign ministry authorized an independent translation of Mr. al-Hilali's remarks, and as the National Post wrote, "predictably, it discovered MEMRI was accurate."

An Australian lawmaker, Michael Danby, who was quoted February 19 as saying, "Hilali's presence in Australia is a mistake," issued a later press release disputing the argument by al-Hilali supporters that his statements had been taken out of context. He wrote: "It is not true that this remark was 'taken out of context.' There is no 'context' for it to be taken out of. Everyone...knows that 'martyrdom operations' is code for terrorist bombings against civilian targets. A shahid is a 'martyr,' one who dies carrying out one of these bombings. During his visit to Lebanon, Sheikh Hilali also gave a clear and unambiguous endorsement to the terrorist organization, Hezbollah. He said that Hezbollah had 'become a model for all mujahideen in the world.' This remark was not taken out of context. Sheikh Hilali's comments were made at a meeting with the infamous head of Hezbollah, Sheikh Nasrallah."

After being exposed as a supporter of terrorist movements and jihad against the West, Mr. al-Hilali made a point of criticizing Al-Qaeda in English. However, during a subsequent interview on Al Arabiya TV on March 9, in Arabic, the mufti reiterated his previous statements, modified to the new circumstances: "When I make Islamic legal statements, I stand behind them, and I fear no rebuke concerning my duty to Allah. Anyone who supports the abhorrent crime that took the lives of thousands of innocent civilians on September 11, 2001, in America understands nothing of his religion, and it is inconceivable that anyone who understands his religion would justify such a barbaric deed...Yes I do support the Palestinian resistance...And I welcome the national and Islamic resistance efforts in Lebanon...

"[This is] a tempest in a teacup, and these statements [against him] have no effect. Australia is a multicultural society, and is not the property of Howard or of the foreign minister. I am an Australian citizen like him. If he thinks the Lebanese Hezbollah, that defended the Lebanese land by means of the Lebanese resistance and with the blessing of the Lebanese, and whose actions have the support of the U.N., of the Secretary Council, of human rights legislation, of revealed religions law and of the man-made laws, if he thinks Hezbollah is a terrorist organization then I disagree with him..."

A further development is a report that an umbrella group of Australian Muslim organizations has been waiting for Mr. al-Hilali to clarify his statements in support of jihad. If they are unhappy with his response, there is talk of stripping him of his title of Grand Mufti.

Top 10 Arab Conspiracy Theories: Part I

May 5, 2004

The head of the Al-Ahram Research Center in Egypt, Abd Al-Munim Said, once said: "We thought that by the end of the 20th century, the Arab mind would be open enough not to explain everything with a 'conspiracy theory'...the biggest problem with conspiracy theories is that they keep us not only from the truth but also from confronting our faults and problems....This way of thinking relates any given problem to external elements, and thus does not [lead] to a rational policy to confront the problem..."

Since September 11, conspiracy theories against America, the Jews, and the Zionists have been rampant in the Arab world. These conspiracies are not only told by marginal personalities and press outlets, but by important members of governments and the press.

Some of last year's most farfetched conspiracy theories in the Arab world include: American soldiers cannibalized Iraqi civilians; America was responsible for the car bomb that killed Iraqi Shia leader Muhammad Bakir Al-Hakim; the Jews were behind the explosion of the space shuttle Columbia, America was behind the SARS virus, and the Iraq war was launched to coincide with the Jewish holiday Purim. The following highlight the top 10 Arab conspiracy theories in recent months:

10. Pakistani Jamaatud-Dawa chief Hafiz Muhammad Saeed was quoted in the Islamic Republic News Agency on November 13, 2003 stating that Al-Qaeda was not responsible for the Riyadh bombing that month, rather, "it is a Jewish and American conspiracy against the mujahideen and Al-Qaeda."

9. According to an editorial in the November 20-23 Yemen Times, the Istanbul bombings during Ramadan this past November could not have been committed by Muslims, as "the international Zionist establishment was keen on instigating this crime. This is strongly supported by the fact that no Muslim in his right mind could ever condone such crimes..."

8. The Islamic Republic News Agency reported on February 28 that America captured Osama bin Laden in a tribal region of Pakistan. It claimed that Defense Secretary Rumsfeld's recent trip to Pakistan was related to the capture. The report said that America will announce the capture shortly before the November presidential elections.

7. A professor at the American University of Cairo, Galal Amin, writing in the April 1-7 edition of Egypt's Al-Ahram Weekly: "There is still doubt that the September [11] attacks were the outcome of Arab and Islamic terror...Many writers...suspect that the attacks were carried out by Americans."

6. Writing in Kuwait's Al-Watan on March 14, columnist Adnan Zayid Al-Kazimi explained the culprit of the Madrid bombings: "I claim with certainty that the ones who attributed all evil to the Arabs and the Muslims are the Zionists, those who are closest to carry out such an operation like the other operations [that they carried out]."

5. Said Al-Subki, columnist for the Saudi daily Al-Watan, also blamed the Madrid bombings on the Jews in its March 19 edition. He criticized Arab intelligence services for being "incapable of discovering the hidden Zionist fingers planning many terror operations in order to entangle the Arabs and Muslims..."

4. The deputy editor of the Egyptian government daily Al-Gumhouriyya wrote an article on March 18 that accused the Jews of perpetrating every terrorist attack throughout the world. Regarding the Madrid bombings that took place March 11, Abd Al-Wahhab Adas, referring to the explosives and cassettes of the Koran found at the site, claimed: "It is obvious that the Jews are the ones who placed these things, in order to prove to the entire world that the Arabs and Muslims are behind the bombings." Mr. Adas added, about the Jews: "It is they who are behind the events of September 11."

3. In an interview with Al-Arabiyya TV, Lebanese Druze leader and parliamentarian Walid Jumblatt said on March 21 that as part of a "born again Christian" scheme which included 9/11, the CIA controls Osama bin Laden.

2. According to the Iranian Mehr News Agency, Hossein Sheikholeslam, the former Iranian ambassador to Syria, said the series of bombings that hit Damascus recently were "a bid to force Iraq's neighbors to submit to their Iraq policy, the U.S. and the Zionist regime orchestrate such terrorist attacks."

He noted, "This is not the first time that the U.S. and Israel have employed Al-Qaeda elements to help them reinforce their terrorist objectives."

1. After the bombing in Yunbu, Saudi Arabia, on May 1, Crown Prince Abdullah said: "Zionism is behind terrorist actions in the kingdom. I can say that I am 95% sure of that."

Top Saudi Cleric to Spread Hate Doctrine in Canada

May 12, 2004

Sheikh Abd Al-Rahman Al-Sudayyis, the Saudi government appointed imam of the Grand Mosque in Mecca, will give a series of lectures in Canada next week and attend the Islamic Society of North America conference in Toronto. Mr. al-Sudayyis's position is one of the most prestigious in Sunni Islam.

His sermons hold significant weight throughout the Islamic world.

The themes of his sermons are characterized by confrontation toward non-Muslims. Mr. al-Sudayyis calls Jews "scum of the earth" and "monkeys and pigs" who should be "annihilated."

Other enemies of Islam, he says, are "worshippers of the cross" and "idol worshipping Hindus" who should be fought. Mr. al-Sudayyis has been consistent in calling for jihad in Kashmir and Chechnya, for Jerusalem to be liberated, and for the "occupiers in Iraq" to also be fought. He often claims that Islam is superior to Western culture.

At the Grand Mosque in Mecca on February 1, 2004, Mr. al-Sudayyis called on Muslims everywhere to unite to defeat the world's occupiers and oppressors.

"History has never known a cause in which our religious principles, historical rights, and past glories are so clearly challenged…The conflict between us and the Jews is one of creed, identity, and existence."

He told those listening to "read history," in order "to know that yesterday's Jews were bad predecessors and today's Jews are worse successors. They are killers of prophets and the scum of the earth. Allah hurled his curses and indignation on them and made them monkeys and pigs and worshippers of tyrants. These are the Jews, a continuous lineage of meanness, cunning, obstinacy, tyranny, evil, and corruption…"

Mr. al-Sudayyis elaborated on the conflict between Muslims and Jews: "O Muslims, the Islamic nation today is at the peak of conflict with the enemies of yesterday, today, and tomorrow, with the grandsons of Bani-Quraydah, al-Nadhir, and Qaynuqa' [Jewish tribes in the early days of Islam]. May Allah's curses follow them until the Day of Judgment."

"The nation must know that these are people with a disgraceful history and a black record... They want to establish the Greater Israel with Jerusalem as its capital. They also aspire to demolish the Al-Aqsa Mosque and build their alleged temple in its place. They want to liquidate the State of Islam and the Koran, and build the State of the Torah and Talmud on its debris. They will get what they deserve from Allah..."

"Our Al-Aqsa is crying out saying all mosques have been liberated, while I — a great holy mosque — am still being desecrated. Is the aspiration of over 1 billion Muslims to preserve their holy places [to be] considered savagery and terrorism? What a great lie, O Allah, O steadfast brothers in struggler and steadfast Palestine, the land of honor, loftiness, sacrifice, jihad, and bravery. The captivity of our Al-Aqsa in the hands of the tyrants makes us sleepless. May Allah please us with its liberation. Victory is coming soon, Allah willing."

"...Here are the flags of victory looming on the horizon. We can smell it. It is crowned by a brave jihad, an intifada, which is still the winning card and the lit candle in the hands of the devout sons of this nation...O nation of jihad and sacrifice, it is the duty of Muslims to support their brothers in creed in Palestine and elsewhere and to back them with material and moral support. Jihad with money sometimes supersedes jihad with soul, as mentioned in many Koranic verses and the prophet's traditions."

In a sermon on April 23, 2004, regarding Iraq, Al-Sudayyis said that "our Muslim brothers in the Iraq of history and civilization are facing another bloody chapter, particularly in the brave, steadfast city of Al-Fallujah."

He called on Muslims everywhere to unite "to defeat all their occupiers and oppressors" for the destruction of the enemies of Islam, to support "our mujahideen brothers in Palestine," and to disperse "the unjust Zionists."

Discussing plots by enemies of Islam, who he identifies as Hindus, Jews, and Christians, Mr. al-Sudayyis delivered a sermon on May 31, 2002, which said: "Those whom Allah cursed, got angry with, and turned into monkeys and pigs, the tyrant worshippers among the Jewish aggressors and criminal Zionists.

"Their course is supported by the advocates of usury and worshippers of the cross, as well as by those who are infatuated with them and influenced by their rotten ideas and poisonous culture among the advocates of secularism and Westernization...

"The enemies of Muslims among the atheists insist on their arrogance and aggression against our people and our holy places in Chechnya? The idol-worshipping Hindus indulge in their open hatred against our brothers and holy places...in Muslim Kashmir, threatening an imminent danger and a fierce war in the whole Indian subcontinent?...O Allah, support our brother mujahideen for your sake and the oppressed everywhere. O Allah, support them in Palestine, Kashmir, and Chechnya. O Allah, we ask you to support our Palestinian brothers in Palestine against the aggressor Jews and usurper Zionists. O Allah, the Jews have oppressed, terrorized, and indulged in tyranny and corruption. O Allah, deal with them for they are within your power."

According to Mr. al-Sudayyis, Islam is superior to Western culture. He told worshippers in Mecca in February 2002: "The most noble civilization ever known to mankind is our Islamic civilization. Today, Western civilization is nothing more than the product of its encounter with our Islamic civilization in Andalusia [medieval Spain]. The reason for [Western civilization's] bankruptcy is its reliance on the materialistic approach, and its detachment from religion and values.

"[This approach] has been one reason for the misery of the human race, for the proliferation of suicide, mental problems...and for moral perversion...Only one nation is capable of resuscitating global civilization, and that is the nation [of Islam]... While the false cultures sink in the swamp of materialism and suffer moral crises...our Islamic nation is the one worthy of grasping the reins of leadership and riding on the back of the horse of pioneering and world sovereignty."

"Read history," Mr. al-Sudayyis said in another sermon in May 2002, "and you will understand that the Jews of yesterday are the evil fathers of the Jews of today, who are evil offspring, infidels, distorters of [others'] words, calf-worshippers, prophet-murderers, prophecy-deniers...the scum of the human race 'whom Allah cursed and turned into apes and pigs...' These are the Jews, an ongoing continuum of deceit, obstinacy, licentiousness, evil, and corruption..."

The concluding supplications of Mr. al-Sudayyis's sermons are often filled with statements concerning current affairs. He consistently calls for "Muslims to humiliate the infidels [non-Muslims], as well as their destruction."

For example, on November 1, 2002, he stated "O Allah, support our mujahideen bothers in Palestine, Kashmir, and Chechnya and destroy the aggressor Jews and the tyrannical Zionists, for they are within your power."

In a June 21, 2002, sermon, Al-Sudayyis gives supplication: "O Allah, support them in Palestine, Kashmir, and Chechnya. O Allah, deal with the Jews and Zionists for they are within Your power. O Allah, scatter their assemblies, make them a lesson for others, and let them and their property be a booty for Muslims."

In another sermon in May 2003, Mr. al-Sudayyis condemned what he termed the "serpents" to "spit their venom" by harming the Islamic religion, ridiculing the pious, and blaming the school curricula and religious and welfare institutions.

Mr. al-Sudayyis said: "O Lord, support our brother mujahideen for your sake everywhere. O Lord, support them in Palestine. O Lord, deal with the aggressor Jews and sinful Zionists. O Lord, deal with them for they are within your power. O Lord, deal with the enemies of religion and show us the miracles of your power on them."

Also, on July 11, 2003, he said: "O Allah, support our mujahideen brothers everywhere. O Allah, help them score victory over the unjust Jews and aggressive Zionists in Palestine. O Allah, destroy the Jews and their supporters. O Allah, destroy them, for they are within your power. O Allah, disperse them and make them prey for Muslims."

According to statements beginning in June 2003 made by Saudi Embassy spokesman Adel Al-Jubeir, "Hundreds of imams [in Saudi Arabia] who violated prohibitions against preaching intolerance have been removed from their positions and more than 1,000 have been suspended and referred to educational programs."

Clearly, this is not the case with Saudi Arabia's leading imam, Sheikh Abd Al-Rahman Al-Sudayyis, who continues to preach incitement from the most holy site in all of Islam.

The History of Extremists' Love of Death

May 19, 2004

In the war on terrorism, major battles from early Islamic history serve as inspiration for those fighting against the West. As Al-Qaeda associate Abu Musab al-Zarqawi beheaded Nicholas Berg in Iraq last week he said, "...Is it not time for you [Muslims] to take the path of jihad?...The Prophet ordered [his army] to strike the necks of some prisoners in [the battle of] Badr and to kill them...And he set a good example for us. As for you Bush, dog of the Christians...You will only get...coffin after coffin slaughtered in this manner."

Another chapter from early Islamic history serving as a lesson for today's Muslims at war against the West is the concept of love of death.This originated at the Battle of Qadisiyya in 636 when the commander of the Muslim forces, Khalid ibn Al-Walid, sent an emissary with a message from Caliph Abu Bakr to Persian commander Khosru. The message said: "You should convert to Islam, and then you will be safe, for if you don't, you should know that I have...an army of men that love death, as you love life." This account is recited in today's Arab sermons, newspapers, and textbooks.

In his speech on March 19, President Bush referred to this concept: "On a tape claiming responsibility for the atrocities in Madrid, a man is heard to say, 'We choose death, while you choose life'...It is a mindset that rejoices in suicide, incites murder, and celebrates every death we mourn. And we who stand on the other side of the line must be equally clear...We do love life...We believe in the values that uphold the dignity of life...and we know that this way of life is worth defending. There is no neutral ground in the fight between civilization and terror..."

Muslim clerics often refer to the love of death.The chief Palestinian Authority cleric, mufti Sheikh Ikrimeh Sabri, said, "We tell them, in as much as you love life — the Muslim loves death and martyrdom...The Muslim loves death and [strives for] martyrdom..."

Saudi Sheikh Abd Al-Muhsin Al-Qassem in Al-Madina added: "The Jews...are cowards in battle...they flee from death and fear fighting...They love life..."

The former head of the Al-Azhar Fatwa Committee, Sheikh Atiyyah Saqr, was asked on March 22 about "the Jews' main characteristics." He answered, "Cowardice and love for this worldly life are undisputable traits [of the Jews]." The Lebanese terrorist organization Hezbollah's secretary general, Hassan Nasrallah, recently said: "We have discovered how to hit the Jews where they are the most vulnerable. The Jews love life, so that is what we shall take away from them. We are going to win, because they love life and we love death."

Abdallah Al-Naggar, a columnist for the Egyptian government daily Al-Gumhuriya, has written about the love of death: "The believers in Allah rightly do not dread their enemies... Their enemies protect their [own] lives, as criminals do...The believers do not fear the enemy [during] the struggle and do not protect their lives... Yet their enemies protect [their] lives like a miser protects his money...This is the secret of the believers' victory over their enemies..."

In a satiric article published in Al-Hayat, liberal Egyptian playwright Ali Salem suggested opening a kindergarten to teach terrorist values: "...You will easily notice that they love life, and that is the weak point that we will exploit.We, in contrast, love death and protect it..." The Tunisian intellectual, Al-Afif Al-Akhdar, asked in an article for the liberal Arabic Web site, www.elaph.com: "Why do expressions of tolerance, moderation...and negotiation horrify us [Muslims]?...Why do other people love life, while we love death, slaughter, and suicide, and [even] call it heroism and martyrdom?..."

As the war on terror continues, the voices celebrating death over life have been heard more often than those criticizing this philosophy. An editorial in the Lebanese Daily Star on May 13 warned: "The region's kings, princes, and presidents need to learn a valuable lesson from this abhorrent incident: that fractured societies produce real-life theaters of shame like the Berg murder ... If the Berg beheading does not catapult the region's leaders from the world of lethargy to the world of vigorous action to establish law and order in their own societies...then they will be considerably weakened...Today...Berg was put to the sword; tomorrow, it could be the Arab nation..."

Arab World Acts on Bush Reforms

May 26, 2004

Since the Bush administration announced the U.S. Middle East Partnership Initiative on December 12, 2002, governments in the Middle East have given this and subsequent American-sponsored reform initiatives a chilly reception. The concept of reform, however, has since taken root in the Arab world in a manner previously unimagined. Even the Arab League grudgingly committed itself to reform this week.

As the president of the American University of Kuwait, Shafeeq Ghabra, wrote in the Lebanese Daily Star on March 16: "The issue of reform was not even addressed in the Arab region a few months ago... It was officially dealt with when the U.S. administration brought it up. Talk of reform will most probably disappear if the U.S. administration stops calls for it... [But] Arab countries that refuse reforms being imposed are the same ones which adopted imported socialist and revolutionary thought and imported the doctrine of military coups, capitalist doctrine, or other ideas generated by the West."

While the governments of the Middle East insist that they will never accept reform dictated from outside, a look at each country reveals that widespread debate regarding reform is now underway. Conferences on reform have been held this year in Yemen, Egypt, Jordan, and Qatar. The daily headlines of most Arabic newspapers discuss reform, while substantial time is given to it on television. Almost every speech made by Arab leaders also mentions reform.

A growing list of Arab states, including Yemen, Morocco, Algeria, Tunisia, and Mauritania are dependent on grants directly from the MEPI. The initiative focuses on four pillars: economic, political, educational, and women's issues. Other Arab states such as Egypt and Jordan increasingly rely on American foreign aid, some of which is directed to reform initiatives.

According to State Department statistics, America quadrupled funding for the MEPI since it was launched. In 2002, $29 million was committed by the Bush administration for the MEPI, while $100 million was spent in 2003 and $145 million for 2004.

The latest MEPI grants were announced on May 7, and involved $5 million to be spent on: strengthening democratic processes in Moroccan communities; establishing a TV network for girls and young women which will be broadcast in Qatar, the UAE, Egypt, and Jordan; building business aptitude for Yemeni youth; assisting women and civil society in Jordan; and bringing young Arab leaders for business training in America.

Recent articles in the Arab press have reported on successful MEPI projects such as e-learning classrooms, health clinics for women, conferences focusing on transparency and accountability in business, and workshops devoted to women's rights. As Arab countries are underdeveloped, the impact of these projects on society has been tremendous.

An increasing number of Arab figures are recognizing the importance of President Bush's reform initiatives, and have come out in support of them. The former Jordanian information minister, Salleh Al-Qallab, wrote in the London Arabic daily Al-Sharq Al-Awsat on March 10, "The greatest lie that the opponents to reform in the framework of the Greater Middle East Initiative are trying to propagate is that all the Arab nations reject the initiative - which is untrue and inaccurate." Even the anti-American columnist Jihad Al-Khazen wrote in Al-Hayat on April 7, "The American suggestions for reform are not bad, despite the intentions of the Bush administration, and they must not be rejected simply because they stemmed out of the U.S."

The United Nations has also embraced the Bush administration's call for reform in the Arab world under the auspices of the United Nations Development Program, with the release of two reports on Arab Human Development. The main author of these reports, Nader Fergany, was accused by many in the Arab world of exposing the dirty laundry of the region: examining the lack of freedom, high illiteracy rate, and low per-capita GDP within the Arab world. Writing in Al-Hayat on March 17, Mr. Fergany claimed, "The American project for the Middle East is based on the AHDR [Arab Human Development Programs].This might add credibility to an Arab serious step toward reforms...."

While Arab leaders have publicly declared their apprehension of American reform initiatives, they have not refused American funds or the implementation of programs devoted to reform within their countries.

Unique Arab Response to Abu Ghräib

June 2, 2004

Responding to Abu Ghräib, several Arab journalists have noted that the actions of what a handful of Americans did pales in comparison to what occurs daily within prisons of the Arab world. Articles in the Arab press have begun calling on Arab leaders to end human rights violations within Arab prisons.

The Egyptian president, Hosni Mubarak, called the Abu Ghräib abuse "abhorrent and sickening, and against all human values and human rights ..." But as Yusri Fatyan wrote in the Egyptian paper Al Arabi on May 23: "Frankly, what happens in Egypt doesn't differ much from what happens in Iraq's prisons ... So that we don't get a surprise when foreign organizations start talking about some of our police stations, like Helwan, Al Sahel, Bilqas and others ..." Another Egyptian journalist, Muhamad Ali Al Farra, wrote in the Islamist paper Al Sha'b: "Some Arab rulers have practiced torture on people which no one would believe, and even finishing with tortured bodies by burning them in acid, so how could such rulers condemn torture of Iraq prisoners? Who is going to throw stones at others when his own house is made of glass?"

While the chairman of the Saudi Shoura Council, Saleh Bin Humaid, called actions in Abu Ghräib "hideous scenes of human rights violations," prisoners in the Kingdom have almost no legal rights, with punishment based on Islamic law including fingers and hands cut off for stealing and beheadings for drug dealing.

The Syrian minister for expatriate affairs, Buthayna Sha'ban, wrote in Al-Sharq Al-Awsat: "The pictures ... of the Abu Ghräib prison aroused revulsion ... because of the extent of the contempt for human dignity and fundamental human rights — particularly on the part of the forces that claimed [they had crossed] the oceans to rescue the Iraqi people from the inhumane actions [by Saddam's regime] and to bring freedom and democracy ... It is the American administration's supercilious view of the Arabs and Muslims ... and the racist campaign against Islam ... that leads to crimes of this kind ..."

The editor-in-chief of the Kuwaiti daily Al-Siyassa, Ahmad Jarallah, responded in an editorial: "Dr. Buthayna Sha'ban, who is 'revolted' ... should be the last to express her revulsion — because the kinds of torture carried out in the prisons of the regime of which she is a part and in whose services she acts are too numerous to count. No atrocity surpasses the kinds of torment and torture [in the Syrian regime] ... We have gone overboard in our talk of the Abu Ghräib torture scandal ... None has dared acknowledge that the U.S. behaved properly in uncovering [this] scandal, for having sufficient courage to apologize. It could have remained silent, or denied it — as is the custom of some Arab regimes that torture, assassinate, bury alive, rip out fingernails, and dissolve [people] in pits of acid, and appear before the world like innocent children with angels' wings ..."

Other Arab journalists have recognized that American actions in Iraq are not based on any sinister plans and that while America has made mistakes, it has done more for Iraqis than any Arab state. On the liberal Arabic Web site www.elaph.com, columnist Hayan Nayouf wrote: "After the scandal of the torture of Iraqi prisoners ... the Arab media handled this affair in a way arousing ridicule, proving that the Arab media and intellectuals possess everything but objectivity, transparency, and disclosure of the truth ... The American president ... apologized for the deeds of the American soldiers, and all the Americans also apologized ... And then the Arab intellectuals came, with their mocking, idiotic, and illogical media, and ridiculed this apology ... The question arises whether Saddam or any other Arab leader [ever] apologized? Did Saddam apologize to the Iraqi people for burying a million Iraqis ... for murdering innocents in his prisons ..."

Since the fall of Saddam's regime much has been revealed of what transpired within his prisons. The treatment of prisoners went far beyond the incidents by Americans of humiliation and beatings that occurred in Abu Ghräib. Iraqi human rights activist Ibrahim Al Idrissi, the president of the Association for Free Prisoners, an Iraqi NGO, has documented the execution of 147,000 political prisoners under Saddam's regime. Mr. Idrissi recounted one incident in which a woman was raped by 12 men, and then her unborn child was cut from her stomach. He told Lebanon's the Daily Star on May 24 that American abuse in Iraqi prisons was a "joke" compared to that under Saddam's regime.

The Creation Of the Berg Conspiracy

June 9, 2004

The beheading of Nicholas Berg in May by terrorist Abu Musab al-Zarqawi inspired the latest in a long line of conspiracy theories that have emerged from the Arab and Iranian press.

Ibrahim Al-Fayoumi, of Egypt's Al-Azhar University, said on May 12 that the beheading of any person, dead or alive, is forbidden by Islamic law, but that he thought that the Berg case was "American propaganda to divert from the [Abu Ghraib] scandal."

An article headlined "The Unexamined Life of Nicholas Berg," which appeared in the Saudi daily Arab News on May 21, questioned Berg's presence in Iraq, alluding to his alleged connections to the Mossad as a "communication spy." The article also placed doubt on Mr. al-Zarqawi's involvement in Berg's murder.

The Iranian daily Tehran Times on May 18 included an article that called the Berg beheading a fraud: "Not only were the purported screams of Nick Berg not in synchrony with the decapitation, but there was also a total lack of blood...We forwarded the video to...a forensic expert... He wrote back...that in his opinion the video is a fraud..." The paper cited La Voz de Aztlan as its source. According to the Southern Poverty Law Center, the L.A.-based paper is "the racist and antisemitic Web magazine of a tiny group of extremist Chicanos."

On May 20, Sheikh Attar, an Iranian political commentator, appeared on Iranian channel Jaam-E-Jam 2 and blamed the CIA for murdering Berg, to overshadow the Abu Ghraib prison scandal.

He said previous American actions had created such a precedent: "A chance observation of the Bay of Pigs and Cuba affair [shows that] a senior CIA official told [CIA director] Allen Dulles, 'We can sacrifice seven or eight Americans there.' That was supposed to serve as pretext for attack, but Kennedy opposed it. This is not unprecedented. Absolutely, that is their method."

A May 15 story on Aljazeera.net discussed conspiracy theories in online Arab chat rooms regarding the Berg beheading, and the article "raised numerous questions concerning its authenticity." Questions involved: why Nick Berg was wearing an orange jumpsuit, the timing of the executioner's announcement that Berg was to die for "Iraqi prisoner abuse," and Berg's Jewish background. Other questions included why "the body is completely motionless even as the knife is brought to bear..." The Arab bloggers write that the accent of the executioner is not Iraqi or Jordanian, and that Mr. al-Zarqawi was believed to have been killed in March. One writer observes that his face is so well-known that "why would he bother to cover it?"

Fares.net, one of the first Arabic Internet sites, posted on May 17 discussions on Berg's killing, with the conclusion being that the videotape was a fabrication. Translations of the Arab bloggers statements include: The bodies of the killers looked husky, unlike the thin and agile bodies of Iraqis. The hands of those people looked very white. Their motions looked like those of Westerners. In the videotape a Western voice was heard saying "thy will be done," which is an expression that even an Arab well versed in English would not use; and the chair on which Berg was sitting is the same kind that Lynndie England, who is accused of abuse in Abu Ghraib, was photographed sitting in, and is the same kind as chairs seen in that prison.

Other questions brought up by the Arab bloggers include: Was Berg used to carry out the bombings that Mr. al-Zarqawi was accused of? Did America set up the killing to save President Bush's standing in the upcoming elections? One person wrote "If you thought that the Americans could not possibly go that far...remember the black history of the CIA...and remember who killed the president of the U.S. John Kennedy?" Another wrote, "Since Berg visited several countries in the world, he could have been a Mossad agent, and the Mossad wanted to get rid of him..."

The site included side-by-side close-up pictures of Berg, one with a beard from the beheading videotape, and another one of him bespectacled, released by his family. A line was drawn along his nose, which was supposedly straight in the family picture and crooked in the other one. The caption under the pictures says: "Notice the difference. The real Berg is on the left, not the one in prison garb."

Like many other conspiracy theories circulating in the Middle East, this one too will likely gain acceptance as time goes by.

Muslim Reformists Vs. Wahhabism

June 16, 2004

An increasing number of reform minded Muslims have begun to speak out against the impact of Saudi Wahhabism in the Muslim world. They have accused Wahhabism of serving as Al-Qaeda's guiding philosophy, "poisoning minds" of young Muslims, and being the main purveyor of anti-American, anti-Semitic, and anti-Christian sentiment in the Arab and Muslim world.

Liberal Egyptian intellectual Tarek Heggy, author of "Culture, Civilization and Humanity," recently wrote about the need for Muslim moderates to work against Wahhabism: "What needs to be done at this stage is to champion the cause of enlightenment by supporting moderates and promoting the humanistic understanding of Islam...Efforts in this direction must go hand in hand with a counter offensive against the rigid, doctrinaire, even bloodthirsty, version of Islam that first appeared among isolated communities separated from the march of civilization by the impenetrable sand dunes of the Arabian Desert."

Mr. Heggy, who will embark on a speaking tour in Washington, D.C., in late June to discuss his new Egyptian think-tank and newspaper, added: "The time has come for the Saudi government to part ways with Wahhabism and to realize that the alliance between the House of Saud and the Wahhabi dynasty is responsible for the spread of obscurantism, dogmatism, and fanaticism, poisoning minds with radical ideas opposed to humanity."

Fellow Egyptian Wael al-Abrashi, deputy editor of the Egyptian weekly Roz Al-Yousef, wrote a series of articles critical of Wahhabism last year following the Riyadh bombings. According to Mr. al-Abrashi: "Saudi Arabia has become the biggest area for extremist ideology and [provided] the broadest scope for the development of its viruses."

Since his articles were first released, Al-Qaeda has attacked Saudi Arabia multiple times, something which Mr. al-Abrashi warned would happen: "Although Saudi Arabia has adopted a strategy of exporting Wahhabism to the rest of the world...Saudi Arabia created the monster, exported it abroad, and then lost control of it. Then, the monster turned on it..."

Ethiopian journalist Alem Zelalem has written extensively about Wahhabism "corrupting" the Islam of his native Ethiopia and described how Saudis have built hundreds of mosques in Ethiopia over the past eight years, popping up "like weeds." One way the Saudis have spread Wahhabism to Ethiopians, he explains, is "by taking advantage of the unfortunate economic conditions of the downtrodden Ethiopian masses, the Saudi Embassy in Addis Ababa is busy bribing people to convert to Islam. The usual amount that they pay...is some $600.00." Mr. Zelalem called the mosques and madrassas "brainwashing factories" for teaching jihad and anti-American, anti-Christian, and anti-Semitic ideology. He detailed how at least 5,000 Ethiopian boys have undergone "military training" for jihad in the Middle East.

In an article in the Addis Tribune about the corruption of Somali Islam by Saudi Arabia, journalist Bashir Goth criticized the Saudis' spread of Wahhabism to his country: "It is a pity to see that, at a time when Saudi Arabia, the home of Wahhabism, is reassessing the damage that Wahhabism and extremism had done to their country's name and to the reputation of Islam all over the world, that Wahhabism has to find a save haven in our country." He also discussed how the Saudi-trained religious police have taken over Muslim towns: "The most conspicuous foot soldiers of Wahhabism are the moral police known as Mutawi'un, who roam in the streets like riot police and force people to perform rituals or adhere to Wahhabism's code of decency...This is the Wahhabism that the Saudi-oriented clerics want to impose on Somaliland...It is a closed [mindset] that turned Islam into a fragile creed that lives in constant fear of children's toys and games such as Barbie dolls and Pokemon..."

Even in Saudi Arabia, the detrimental effects of Wahhabism are now discussed in public. Jamal Khashoggi, who currently advises Saudi Prince Turki in London, was fired last year following the Riyadh bombing when the paper which he edited, Al–Watan, included articles critical of Wahhabism's spiritual father, Ibn Taymiyya. One article written by Khaled al-Ghanami condemned the Saudi government's religious police and criticized the "spiritual father of Wahhabism," calling his philosophy "the real problem," and "a mistake" for Saudi Arabia.

As current events play out in Saudi Arabia, the royal family would do well to listen to the critique by reformist Muslims on how Wahhabism has negatively impacted their communities. As Waelal-Abrashi explained, Saudi Arabia created the Wahhabi monster — and then lost control of it.

They must now figure out how to battle it.

Must-See Saudi TV

June 23, 2004

The Saudi ambassador to America, Prince Bandar bin Sultan, said in a June 15 press release that "senior religious scholars in Saudi Arabia have continually and unequivocally condemned terrorism. In our war against terrorism, these condemnations are a powerful weapon."

In this statement, as well as in many others by the Saudi Embassy in Washington since the Riyadh bombings on March 12, 2003, the government of Saudi Arabia encouraged a message of tolerance. As the director of the Saudi information office, Nail al-Jubeir, said on May 6, 2004: "Our religious establishment has taught moderation to our people. The religious establishment has demonstrated that it is the body most competent and effective at delegitimizing Al-Qaeda's cultist ideology."

As part of MEMRI's TV Monitoring Project, Saudi government-controlled television channels including TV1 and TV2, and satellite channels, such as Iqraa TV, are continually monitored. These channels include shows with leading Saudi religious figures, professors, members of the royal family, government leaders, and intellectuals. Constant themes within Saudi television shows include: calls for the annihilation of Christian and Jews, rampant anti-Americanism and anti-Semitism, and support for jihad and incitement against American troops in Iraq. Segments from these TV shows can be found at www.memriTV.org.

A professor at Um Al-Qura University in Saudi Arabia, Sheikh Ahmad Abd Al-Latif, was asked the following question on Saudi channel TV1 on May 24: "Some imams and preachers call for Allah to annihilate the Jews and those who help them, and the Christians and those who support them...Is it permitted according to Islamic law?" Professor Al-Latif responded: "What made them curse the Jews is that the Jews are oppressors...The same goes for the Christians, because of their cruel aggression against Islamic countries...while the truth is that this is a crusading war whose goal is to harm Muslims. This is why a Muslim is allowed to curse the oppressors from among the Jews and Christians...Cursing the oppressing Jews and the oppressing and plundering Christians and the prayer that Allah will annihilate them is permitted."

Sheikh Muhammad Al-Munajid, a disciple of one of Saudi Arabia's most revered religious leaders, Sheikh Abd Al-Aziz ibn Abdallah ibn Baaz, was identified in a report in the Washington Post on December 11, 2003, as running "a Web site that promotes intolerance of Christians and Jews and calls for holy war on Shiite Muslims" and was included as one of 16 clerics associated with the Saudi Embassy's Islamic Affairs Department who was stripped of diplomatic credentials.

Mr. al-Munajid said on Iqraa TV on April 15: "The issue is not one person, two, ten or a hundred going out with their guns to support their brothers. Defeating the infidels requires a much greater effort. It requires the mobilization of the nation. How can the nation be mobilized? I believe that the stupid acts of these Jews and Crusaders mobilize the nation. The big explosion will come! In spite of everything, it will happen!"

Much of Saudi TV is based upon religious programming. Many of these programs refer to the spread of Islam throughout the world and the battle against non-Muslims. On a May 20 episode of Iqraa TV's "Mushkilat Min Al-Hayat," or, "Problems from Life," Saudi Sheikh Abdallah al-Muslih, chairman of the Commission on Scientific Signs in the Koran and Sunnah of the Muslim World League, used evidence from early Islam to support his claim that suicide bombings on enemy land are permitted according to Islamic law: "Regarding a person who blows himself up, I know this issue is under disagreement among modern clerics and jurisprudents... there is nothing wrong with (martyrdom) if they cause great damage to the enemy. We can say that if it causes great damage to the enemy, this operation is a good thing. This is when we talk of Dar Al-Harb [House of War]. But, if we speak of what happens in Muslim countries, such as Saudi Arabia...this is forbidden, brothers! This is the land of the Muslims. We must never do this in a Muslim country."

Supporting jihad against American troops in Iraq is the topic of many Saudi TV programs. On May 10, Yassin al-Khatib, a professor of Islamic law at Um Al-Qura University, declared on Al-Majd TV that "the honor, blood, property and mostly the fact that they entered the country (i.e. Iraq)... make it every Muslim's duty to go out against them, not only the Iraqis. This is every Muslim's duty. Jihad today has become an individual duty that applies to each and every Muslim. It is forbidden for a person to remain silent... When the Muslims fought in Afghanistan they destroyed the Soviet Union, which was a superpower. It collapsed and Allah willing, so will this (America) collapse."

On a March 17 broadcast on Iqraa TV Saudi preacher Sheikh Said al-Qahtani explained in what cases Muslims are permitted to declare a defensive jihad: "We did not occupy the U.S., with 8 million Muslims, using bombings. Had we been patient, and let time take its course, instead of the 8 million, there could have been 80 million (Muslims) and 50 years later perhaps all the U.S. would have become Muslim...What should a Muslim do if he is attacked in his country, on his land? In this case, there is no choice besides defense, self sacrifice, and what religious scholars call defensive jihad…We attacked their country, and this caused them to wake the dormant enmity in their hearts…Especially since there is global Zionism, the enemy of Islam, and Judaism, and fundamentalist Crusaders…They interpret this whole incident as only the beginning and thus there is no choice but a pre-emptive strike."

Mr. al-Qahtani added on another Iqraa TV show on May 5: "Allah said, 'prepare against them all the force and horsemen that you can.' What for? In order to strike fear into their hearts…At the same time, (we should) establish strategies for the future, even if only for the short term, and prepare…so that one of these days, even 100, 200, or 400 years from now, we will become a force that will be feared by the infidel states."

One month prior to Canada's decision to bar Saudi Arabia's highest-ranking government-appointed cleric, Sheikh Abd Al-Rahman Al-Sudayyis, from entering the country, the Saudi preacher gave a sermon on Saudi channel TV 1 discussing jihad. Mr. al-Sudayyis, who serves as one of the imams of the Grand Mosque in Mecca, stated: "The history of the Jewish people is written in black ink, and has included a series of murders of the prophets, the mujahedeen, and righteous people… But maybe it is the beginning of their end…You have revived the hopes of this nation through your blessed jihad… With Allah's help, one of two good things will be awarded you: either victory or martyrdom."

On June 11, Mr. al-Sudayyis was given the honor of leading Friday prayers when over 55,000 worshippers gathered in the East London Mosque to hear him call, in English, for interfaith peace and harmony in an event that included Prince Charles via satellite and Rabbi Jonathan Sacks, the chief rabbi of England.

In reaction to the May 1, 2004, terrorist shooting in the offices of an oil contractor in Yunbu, Saudi Arabia, in which seven, among them two Americans, were killed, Saudi Crown Prince Abdallah ibn Abd al-Aziz said at a gathering of Saudi dignitaries, including top Muslim clerics and preachers, that "the Zionists" were to blame. The statement was made at a May 2 gathering of the royal family, which aired on Saudi TV 1: "Be assured that the Zionists are behind everything. This is certain. I don't say this with 100% [confidence], but with 95%."

In addition to other leading Saudi officials, such as the interior minister, Prince Nayef, the following day, Salem al-Aufi, the secretary general of the King Fahd Center for Printing the Koran, reiterated on Saudi TV 1: "Yesterday we heard the crown prince say that Zionism is behind these attacks. Who helps them carry out these kinds of attacks? Who gives them money? Who gives them weapons? Who presents them with the evil ideology that brought them to this? There is no doubt that behind this, there are international organizations aspiring to undermine this country's security, and terminate its achievements, and at the top of these organizations stands Zionism, the enemy of the religion."

According to Web sites of the television channels mentioned in this article, programming is presented by the Saudi Ministry of Culture and Information and is advertised as a "balanced blend of religious and cultural programs." The Ministry of Culture and Information's Web site indicates that it is responsible for the kingdom's television and radio broadcasting and publication of printed material, and one of its defined roles includes the undertaking of information campaigns for the kingdom. Therefore, the Saudi family can end its support for jihad programming at any time.

Beheadings Wave Inspired By Saudis

June 30, 2004

In recent years, the act of beheading has become a popular trend among Islamist terrorists, who cite the Koran's Sura Muhammad, Verse 4, as its theological basis: "When you meet in battle those who have disbelieved, smite their necks; and after the slaughter fasten tight the bonds, until the war lays aside its burdens. Then either release them as a favor, or in return for ransom."

"Hitting the neck," or, in Arabic, "Darb Al-A'nak", literally speaking, is one of several words in Arabic for killing. It is not "murdering," since the word "murdering" carries a derogatory meaning. It is simply "killing." This word, which is taken from ancient Muslim vocabulary, preserves the actual manner in which beheadings have been carried out. Therefore, beheading is not seen as unusual or inhumane. It is an act of Islamic punishment under sharia, or, Islamic law.

The beheading of Daniel Pearl in February 2002, followed by the killings of Nicholas Berg, Paul Johnson Jr., and Kim Sun-Il, and the kidnappings and threats to behead a growing list of captives, have garnered major press attention for the terrorists as Arab satellite channels rush to air these acts.

The recent killings of Westerners have been condemned by the Saudi royal family. The Saudi Press Agency reported that the vice governor of the Riyadh region, Prince Sattam Bin Abdulaziz, expressed condolences on behalf of King Fahd Bin Abdul-Aziz, to the wife of Paul Johnson. Prince Sattam said that Johnson's beheading was a malicious crime that is rejected by Islam, declaring that "this has happened for the first time in the Kingdom."

However, the Saudi government has carried out beheadings as a form of execution for breaking an assortment of laws. Examples posted by the Saudi Press Agency include: In August 2003, a Sudanese and a Chadian were beheaded for smuggling drugs into Saudi Arabia. An Afghan was beheaded for peddling opium in September 2003. A Pakistani and an Iraqi were beheaded for smuggling heroin into the Kingdom in April of 2003. In October 2002, a Palestinian was beheaded for murdering his father after an argument. Two Pakistanis and an Indian were beheaded in the summer of 2002 for smuggling heroin into Saudi Arabia.

The London Arabic daily Al Hayat reported on May 3 that the punishment of beheading was about to be applied to the son of Prince Nayef Ibn Abd Al-Aziz, Saudi Arabia's interior minister, who murdered another Saudi citizen. In the end, he was pardoned by the victim's father. The life of the killer, Prince Fahd, was spared on May 1 when "[Fahd] was brought to the prison yard, his eyes blindfolded, and the executioner was prepared with his sword to sever his neck...In front of a large throng of citizens, Prince Fahd begged the father of the victim: 'Save my life'... Suddenly, the father of the victim kneeled twice in front of Prince Fahd, who was lying down in front of the executioner... [He] got up and to the cheers of the crowd pardoned Prince Fahd and untied his hands."

Earlier this year, an Al-Qaeda member in Saudi Arabia warned of beheadings in his last will and testament. Hazem Al-Kashmiri, the son of a retired leading general of the Saudi internal security force, informed the world that Al-Qaeda had "a message to the American soldiers...We promise that we will not let you live safely, and you will not see from us anything else, just bombs, fire, destroying homes, cutting your heads..."

The 18th issue of the Al-Qaeda-identified journal Sawt Al-Jihad included an interview with Fawwaz bin Muhammad Al-Nashami, commander of the Al-Quds Brigade, which took responsibility for the May 29 attack at Khobar, Saudi Arabia, in which 22 people were killed. Al-Nashami detailed how Al-Qaeda members cut the throats of non-Muslims, and in one case, beheaded one of their victims — all as "acts of devotion to Allah:" "We turned to the third site...found a Swedish infidel. Brother Nimr cut off his head, and put it at the gate [of the building] so that it would be seen by all those entering and exiting."

In what appeared to be a single case, the beheading of Daniel Pearl has now influenced other Islamists in Saudi Arabia and Iraq. Even in London, the Web site of Islamist Sheikh Abu Hamza Al-Masri features video of children pretending to behead other children. Islamic history includes periods in which beheadings against infidels were a common practice, and it seems that this cycle is about to repeat itself.

Documenting Anti-Semitism On Arab and Iranian TV

July 7, 2004

The annual Cairo Television and Radio Festival held a few weeks ago included submissions from throughout the Arab world such as the Syrian-produced Hezbollah TV Ramadan series "Al-Shatat." The notorious series, which includes blood libels, was denounced by Western governments earlier this year, but is not any more anti-Semitic than what appears daily on Arab and Iranian television.

MEMRI's TV Monitoring Project, www.memritv.org, which has been systematically reviewing Arab and Iranian television over the past three months, has so far found the following recurring themes regarding Jews: Calls for their annihilation; classic Western anti-Semitic motifs such as support for and denial of the Holocaust; and conspiracy theories, such as a Syrian national TV show on June 16 that claimed President Bush is Jewish. Guests include prominent political, religious, academic, and press personalities.

There are four main elements of Arab anti-Semitism and all can be found on the region's television stations. The first is based upon early Islamic roots associated with the Koran, Muslim tradition, and classic Muslim literature. These date back to Muhammad's confrontation with the Jewish tribes of the Arabian Peninsula. Many examples from Muslim tradition call for the annihilation of the Jews. For example, on May 14, Sheikh Ahmad Omar Hashem, former president of Al-Azhar University, delivered the Friday sermon broadcast by Egyptian Television's Channel 1: "You will fight the Jews until the Jew hides behind a stone or a tree, [but] the stone or the tree will say: 'Oh servant of Allah, oh Muslim, there is a Jew behind me, come and kill him.'"

The second element of Arab anti-Semitism is influenced by European and Christian tradition, including blood libels, "The Protocols of the Elders of Zion," and distortions of the Talmud. Examples include the United Arab Emirates religious channel, Al-Majd TV, which broadcast Sheikh Hamza Mansour, secretary-general of the Islamic Action Front Party in Jordan, on March 22, detailing how the Jews "make matza out of innocent children's blood."

Another example is Hezbollah's Al-Manar, which hosted the mufti of Tripoli, Sheikh Taha Al-Sabonji, and the secretary-general of the Islamic Universities Association, Jafar Abd Al-Salim, for a discussion on April 22 about Jewish history. The speakers cited "The Protocols of the Elders of Zion," which detailed how the Jews "spread corruption amongst all human societies...All the pornographic films in the world are made by Jewish companies, and the same goes for drugs..."

The third element of Arab anti-Semitism relates to denying the Holocaust — while at the same time approving of it, and equating Jews/Zionists with Nazism, while borrowing from Nazi propaganda. An example includes a May 20 Palestinian Authority TV discussion between two Palestinian historians, Jareer Al-Qidwa and Issam Sisalem, who called the Nazi atrocities against the Jews "a great lie." Holocaust denier Roger Garoudy was celebrated in an April 16 sermon by Sheikh Ibrahim Mudeiris, which aired on Palestinian Authority TV. During the sermon, Sheikh Mudeiris also praised "The Protocols of the Elders of Zion."

Another example is from a May 27 Al-Manar discussion between Ghazi Rababa, a professor of political science at Jordan University, and Muhsen Al-Awaj from Saudi Arabia, spokesman for the World Campaign for Resisting Aggression. Mr. Rababa mentioned an alleged declaration by President Jefferson calling on the American people to expel the Jews. This fictitious declaration was also used in Nazi propaganda. Mr. Rababa said: "President Jefferson told the American people in an official speech, 'If you do not expel the Jews from your land, they will enslave you and turn you into slaves in your own lands.'"

The fourth element of Arab anti-Semitism is the continuing demonization of Jews, who are accused of being responsible for or manipulating all disasters in modern history. A favorite conspiracy theory now on Arab and Iranian TV is to blame the Jews for 9/11. A former dean of humanities at Cairo's Ein Shams University, Mustafa Shaka, was interviewed by the Saudi-based Islamic channel Iqra TV on June 16, and explained the Jewish connection to the attacks. Jaam-E-Jam 1, the Iranian government TV channel directed at Europe, also broadcast a series about 9/11 on June 1 that claimed Jewish and Israeli involvement in the attack and as proof cited "the Jewish Protocols."

Every example mentioned in this article can be viewed at www.memritv.org, and additional TV segments are added daily. It should be mentioned that the examples cited in this article are not the exception — they are common.

P.A. Sheikh's Message To America

July 14, 2004

Sheikh Ibrahim Mudeiris has become the Palestinian Authority's leading preacher for Friday sermons in Gaza's main mosque, which are broadcast on P.A. TV. To view segments, visit the newly-launched www.memritv.org.

The sermons of Mr. Mudeiris, a paid employee of the Palestinian Authority's Ministry of Awqaf, inspires the Palestinian people each week - often citing battles from early Islamic history as a precedent for today's jihad against America and the Jews.

Anti-American themes of Mr. Mudeiris's sermons often revolve around conspiracy theories. On June 25, 2004, the sheikh blamed America and Israel for the recent terrorist attacks in Saudi Arabia: "Zionism and the U.S. are responsible for what is happening in Saudi Arabia. Yes, we say this without hesitation. America is responsible for what is currently happening in Saudi Arabia...They have a schedule and a plan; each country will be attacked in turn."

A period of Islamic history in which contemporary Muslims can model their jihad against America, according to the sheikh, is the Byzantine period. On May 21, 2004, he compared America and President Bush to a king from the Byzantine Empire: "[The Khalifa Al-Mutasem] used his army and sent a letter to the Byzantine king. He wrote, demonstrating the power of Islam, 'From the head of the believers, Al-Mutasem, to the Byzantine dog'...Why is there no leader, why is there no ruler, why is no one man enough to send such a message to the American dog, to the Byzantine dog of our days, to the oppressor dog...To the Byzantine dog, you son of a bitch, I have brought an enormous army upon you..."

Citing the Koran in an earlier sermon on September 5, 2003, Mr. Mudeiris explained his thoughts on who is the incarnation of the Byzantium Empire: "If we go back in the time tunnel 1,400 years, we will find that history repeats itself... Byzantium represents America in the west...America will collapse, as Byzantium collapsed in the west...The prophet [Muhammad] could...defeat the America of that time, as we will defeat America...Indeed, we consider America to be our no. 1 enemy...and we see it as our no. 1 enemy as long as we learn from the lessons of the

Battle of Tabouk: 'Make ready for them whatever you can of armed strength and of mounted pickets' [Koran 8:60]."

On another occasion, on March 21, 2003, Mr. Mudeiris called President Bush a "dwarf" and "Pharaoh" and stated that America would be destroyed: "...Allah drowns the pharaohs of every generation. Allah will drown the little pharaoh, the dwarf, the pharaoh of all times, of our time, the American president. Allah will drown America in our seas, in our skies, in our land...The crusader, Zionist America has started an attack against our Iraq...If Iraq is defeated, if the nation [of Islam] is defeated in Iraq - this will be our last breath of life...It is not I who say this, it was the little pharaoh [Bush] who announced it when he stated that this was a crusader attack...You all know about the slap in the face which Iraq gave America at the time of the first invasion, when Iraq gave a knockout [punch] to America, the result of which was that pharaoh the father [the elder George Bush] lost the presidency. This is a mark of shame on this family and therefore pharaoh the son returned to power in order to take revenge...This is also a religious war...America will be destroyed, Allah willing..."

As most sermons from the Middle East end with political messages, Mr. Mudeiris usually dedicates supplications to calls for the killing of Jews and for Allah to destroy America. For example, on April 30, 2004, he exclaimed: "O God, support our kinfolk in Iraq. O God, direct the shots of the mujahadeen for your sake. O God, shake the land under the pillars of the United States."

The previous year, on June 16, 2003, Mr. Mudeiris said: "O God, punish our enemies. O God, destroy those who have harmed us. O God, destroy those who harm Muslims. O God, destroy the Jews and their supporters. O God, destroy the United States and its allies." And on March 17, 2003, Mr. Mudeiris concluded his prayer: "O Lord, support us against our enemies. O Lord, destroy our enemies. O Lord, destroy America and Israel. O Lord, shake the land under their feet."

Arab and Iranian Press Weigh In on America's Election

July 28, 2004

The Arab and Iranian press have increasingly discussed the upcoming American presidential election. Most journalists are pulling for a Kerry victory. Other commentators have said that the choice between the two is like "choosing between cholera and the plague."

The Iranian press has been particularly active in monitoring the elections. The editor of the Egyptian Al-Gil newspaper, Nagi Al Shihabi, was interviewed by the Iranian Al-Alam TV on June 13, and said: "The U.S. wants to eradicate our religious and Islamic identity. Bush declared a Crusader war following the events of 9/11...The U.S. established its country over the body parts of 120 million Indians. We must first define the enemy. The no. 1 enemy of the Arab and Islamic peoples is the U.S. and not only Israel...Bush, Allah willing, will go to hell in this coming November."

During his Friday sermon in the city of Qom, broadcast by Channel 2 of Iranian TV on June 19, preacher Ayatollah Javadi-Amoli called on Americans to vote against Mr. Bush: "We advise the people of America not to continue to tolerate this oppressing, ignorant, pillaging, criminal, and discriminating administration. In the future, do not vote for Bush and his ilk."

Writing in the Tehran Times on June 17, Kian Nader Mokhtari praised Senator Kerry, and hoped that he would triumph over the "neo-Nazis " of the Bush administration: "It has long been a tradition in U.S. politics for irresponsible gun-toting Republican presidents to pass on unresolved dilemmas to their Democratic replacements, and Bush will be no exception...Kerry is exactly what the U.S. needs... Kerry's sensible and methodical approach will no doubt go some distance in solving the stinking heap of a mess left over by Bush and his neo-conservatives — for a minute there I was going to type neo-Nazis !...He [Kerry] may be remembered as the president whose decisions saved the U.S..."

The Lebanese minister of information, Michael Samaha, also discussed the election on Syrian TV on March 30: "The most important thing that could happen in the coming American elections if Kerry wins is that the neo-conservatives leave. They have a complete ideology regarding their treatment of the world and specifically the Middle East and the Near East to which we belong. If they leave, the real American America will return."

In an article in Al-Ahram weekly on April 18, an Arab-Israeli member of Knesset, Azmi Bishara, called on Arab-Americans to "rethink their alliance with Bush" and "shift their votes from Bush to Kerry."

But not everyone in the Arab and Iranian press supports Mr. Kerry. In an interview with Al-Siyassah daily, Sheikh Saud Al Nasser Al Sabah, who was Kuwait's ambassador to America from 1981 until 1992, said he hopes Republicans will stay in power: "If the American administration changes in November, it will be catastrophic ...because those Democrats do not understand a thing about foreign policy, and they lack the determination to make decisions the way Bush made them in Iraq and elsewhere." He added, "Our only hope is that this administration will continue for the next few years..."

California State University professor Asad Abu Khalil was interviewed by the Iranian channel Al-Alam TV on June 23. The professor said: "I say that decision-making in American foreign policy is done by more than just one group. There is, of course, the neo-conservative group, but there is another group which is more important and that is the Christian fundamentalists — the Christian Taliban, which is influential in the U.S., to which the American president belongs...The nature of the American empire may change under John Kerry, but we shouldn't do what some PLO representatives in Western capitals did, when they opened champagne bottles when the Labor party replaced the Likud..."

The Saudi-owned Iqra TV hosted a political debate on March 24. The moderator asked the panel their opinions on voting for Mr. Kerry or Mr. Bush, "Sir, do you believe the American voter will vote for Kerry as a reaction to the position of Bush junior, who always supports Sharon?" A Palestinian politician and writer, Abd Al-Qader Yassin, said: "We need to choose between cholera and the plague."

As November approaches, it is interesting to note that four years ago the Arab and Iranian press was giddy about a Governor Bush victory over Vice President Gore, which was seen as an end to the Clinton administration's Middle East policy. The same press is again calling for a change in the White House.

Al Jazeera Has Hit the Mainstream

August 4, 2004

Now with a reported 35 million viewers worldwide, the Arabic TV channel Al Jazeera appears on the surface to have reached mainstream status.

Proof of this is in the network's coverage of the Democratic and Republican national conventions this summer, as well as with its recently released code of ethics pledging to "address every issue or story with due attention to present a clear, factual, and accurate picture."

However, controversy regarding the incitement to violence surrounds the channel, which was launched in 1996. Other issues include dozens of Al Jazeera correspondents having been arrested for connection to terrorist activities. In December 2001 Al Jazeera cameraman Sami Muhyi Al-Din was one of the first, as American forces arrested and took him to Guantanamo Bay. On September 5, 2003, Al Jazeera correspondent Taysir Alouni was arrested by Spanish authorities and charged with "furnishing Al-Qaeda with funds in Afghanistan." Mr. Alouni is infamous for his interview of Osama bin Laden shortly after September 11. More recently, on April 4, 2004, Al Jazeera correspondent Dib Abu Zayed was indicted in an Israeli military court and accused of serving as a liaison for Fatah in Lebanon and supplying Palestinian terrorists with money and weapons.

Other areas in which Al Jazeera could be considered helpful to terrorists is by their airing of exclusive videos provided to them in order to get their messages out — such as showing a beheading and warning of or taking credit for terrorist attacks. It has been argued that the terrorists reciprocate to the network by providing tips as to where and when to film upcoming attacks.

Such actions have been common in Iraq. For example, the Iraqi daily Al-Sabah reported on November 9, 2003, that Al Jazeera correspondent Sattar Karim acknowledged that his office was used to coordinate attacks against coalition forces. American forces have also arrested Al Jazeera cameramen Salah Hasan and Samir Hamza while filming the aftermath of an attack on a police station. Correspondent Anwar Bahjat was also arrested while covering an attack in Baghdad. According to Al Jazeera's Web site, 10 of its employees had been arrested in Iraq as of October 2003.

In response to ongoing actions in Iraq, on July 3 Iraqi intellectuals appealed to the government to expel Al Jazeera from the country because it is engaged "in a hysterical campaign" against Iraq. An editorial in the Iraqi daily Al-Ittihad on April 12 said that Al Jazeera feeds itself "upon the blood of the Iraqis." Iraq's foreign minister, Hoshyar Zebari, accused Al Jazeera on July 27 of inciting violence in his country and threatened to close its Baghdad offices, while the Governing Council in Baghdad banned Al Jazeera from entering its offices and attending its news conferences last year.

The Arab press has also reported that Saddam Hussein had made payments to Al Jazeera before his ouster. It was implied that such action was in exchange for coverage to be positive of his regime and anti-American. One example appeared in the Kuwaiti daily Al-Qabas on December 29, 2003, which reported that Saddam provided financial support to Al Jazeera headquarters in Qatar. Days later, the Iraqi daily Al-Shira coined the phrase "the Society of Saddam's Orphans," which includes a number of prominent Arab journalists and Al Jazeera talk show hosts and producers who benefited greatly from Saddam's bank account. Another Iraqi daily, Al-Mu'tamar, published documents on April 16 connecting Faisal Al-Qassim, one of the leading personalities on Al Jazeera, with Saddam's former intelligence services. The German weekly Der Spiegel claims to have original documents identifying 24 Al Jazeera and press personalities who were paid bribes by the Iraqi mukhabart. The list includes the director of Al Jazeera, Muhammad Jassim Al-Ali, who has since been fired.

While Al Jazeera would like to now be known as a mainstream network, in America it has a poor reputation in the government and business community. This began days before the attacks of September 11, 2001, when FBI agents seized servers hosting the Al Jazeera Web site from a Dallas company. After its Web site was later hacked multiple times, the network's search for a new host was documented in April 2003.

Also, in March 2003 two Al-Jazeera reporters were banned from the New York Stock Exchange, because, according to the NYSE, the exchange is only open to those focusing "on responsible business coverage." Most recently, Al Jazeera was quick to announce its on-location coverage of both political conventions this summer, but according to reports, it was required to remove its logo at the Democratic convention in Boston.

Iranian Talk Of an Attack On America

August 18, 2004

There are growing indications that Iran may be planning an attack on American soil. These indicators are not secret — they appear in speeches, newspaper articles, TV programs, and sermons in Iran by figures linked to the supreme leader of Iran, Ayatollah Ali Khamenei, and other government officials, all discussing potential Iranian attacks on America, which will subsequently lead to its destruction.

A report on May 28 in Al-Sharq Al-Awsat reported that an Iranian intelligence unit has established a center called "The Brigades of the Shahids of the Global Islamic Awakening." The paper claimed that it had obtained a tape with a speech by Hassan Abbassi, a Revolutionary Guards intelligence theoretician who teaches at Al-Hussein University. In the tape, Mr. Abbassi spoke of Tehran's secret plans, which include "a strategy drawn up for the destruction of Anglo-Saxon civilization." In order to accomplish this, he explained, "There are 29 sensitive sites in the U.S. and in the West. We have already spied on these sites and we know how we are going to attack them."

It was reported that America expelled two Iranian security guards employed by Tehran's U.N. offices on June 29, after the mission was repeatedly warned against allowing its guards to videotape bridges, the Statue of Liberty, and New York's subway system. This was the third time the Iranians have been caught in such activities, which could be connected to the sites mentioned in potential plans to attack America.

Mr. Abbassi's speech further detailed that "[Iran's] missiles are now ready to strike at their civilization, and as soon as the instructions arrive from Leader [Ali Khamenei], we will launch our missiles at their cities and installations." In fact over the past few months, Mr. Khamenei has been vocal about the impending "destruction of the U.S." In May, he was quoted in the Iranian paper Jomhouri-Ye Eslami as saying that "the world will witness the annihilation of this arrogant regime." On July 5, in front of a crowd chanting, "Woe to the enemy if Khamenei commands me to wage jihad," Mr. Khamenei said, "If someone harms our people and invades our country, we will endanger his interests anywhere in the world."

Other Iranian religious leaders have also called for the destruction of America. The secretary general of the Guardian Council, Ayatollah Ahmad Jannati, appeared on Iran's Channel 1 TV on June 4 and said, "Every Muslim and every honorable man who is not a Muslim must stand against the Americans, English, and Israelis, and endanger their interests wherever they may be." When he added "They must not have security," thousands in the audience repeated chants of "Islam is victorious, America will be annihilated." On June 25, Mr. Jannati also led prayers and promised, "Anyone who confronted the revolution, the Imam [Khomeini], and our dedicated people eventually collapsed. America is the last one, and Allah willing it will collapse..." The following week Ayatollah Mohammad Emami-Kashani delivered the Friday sermon live on Channel 1, saying America will collapse like Genghis Khan's empire, "I say to you the American people...you will collapse, America will collapse."

"Time bombs within America" is how Iranian lawmaker Hamid-Reza Katoziyan described Muslims within America, who could be behind future terrorist attacks here. Speaking on Iranian TV channel Jaam-E-Jam 2 on July 27, Mr. Katoziyan warned: "The whole group of people belonging to the Arab community and... Muslims living in the U.S. are currently, in my opinion, in a special situation. Perhaps they do not walk the streets with weapons in their hands or attach bombs to themselves in order to carry out a suicide operation, but the thought is there."

Just as statements from Iranian religious and political leaders, as well as TV programs, have focused on attacking America, so has the print press. An editorial in the July 6 edition of the Iranian daily Kayhan, the conservative paper affiliated with Mr. Khamenei, issued another warning for the future: "...the White House's 80 years of exclusive rule are likely to become 80 seconds of hell that will burn to ashes...That very day, those who resist [Iran] will be struck from directions they never expected. The heartbeat of the crisis is undoubtedly [dictated by] the hand of Iran."

The report by the National Commission on Terrorist Attacks Upon the United States asks: "September 11 was a day of unprecedented shock...The nation was unprepared...How can we avoid such tragedy again?" Taking seriously Iran's threatening words and actions is an important first step.

Saudis Peddle Hate, Despite P.R. Campaign

August 25, 2004

The September 11 commission report notes that Saudi Arabia is "a problematic ally in combating Islamic extremism" and that "significant problems remained" regarding its role in the war against terror.

On August 18, the Saudi embassy in Washington, D.C., announced its newest ad campaign. Embassy spokesman Nail Al-Jubeir explained, "We want to put to rest the charges that have been lingering around since 9/11 that the Saudi government or Saudi officials have funded terrorism or extremism. I think the 9/11 report from the Commission vindicated us on this, and we want to make sure that people in this country are aware of it."

This is not the first time the Saudis have initiated a public relations campaign to highlight their efforts against the extremist Wahhabi ideology they espouse. Past efforts failed because leading Saudi officials were quoted making contradictory statements in English and in Arabic.

Examples include high-ranking members of the royal family blaming terrorist acts such as September 11 on "Zionists." The Saudi interior minister, Prince Nayef, declared in 2002: "Who committed the events of September 11...I think they [the Zionists] are behind these events..." According to NBC Nightly News, his statement "completely undercut a $10 million campaign advertising Saudi Arabia's commitment to the war on terror..."

Just last week on Saudi government TV, Saudi cleric Sa'd bin Abdallah Al-Breik spoke on Channel 1 in support of Prince Nayef's theory: "We must not inflate [the importance] of Al-Qaeda, to the point of claiming that it is the main and only perpetrator of this large operation [9/11]... It is a mistake to ignore the possibility that the Zionist hands used some people who were planted into one of the stages of this plan...These false accusations and the rush to accuse Saudi Arabia, the judging of others according to the guidance of the Zionists via the media which is owned by the Zionists..."

A more recent PR flap occurred in April, a week after a group of leading Saudi royal family members, led by the foreign minister, Prince Saud Al-Faysal, made their way to American thinktanks and TV shows with a message to the American people that they are partners in the war on terror.

The following week, which included a terrorist attack in Saudi Arabia, Crown Prince Abdullah stated at a gathering of Saudi dignitaries that "Zionism is behind terrorist actions in the kingdom ...I am 95% sure." His statement was affirmed by leading Saudis such as Prince Nayef and the governor of the Riyadh region, Prince Salman. Saudi Arabia's foreign policy adviser, Adel Al-Jubeir, appeared on CNN, but was at a loss for words to explain this statement. Rep. Tom Lantos, a Democrat of California and the ranking member on the House International Relations Committee, reacted on the House floor, calling Prince Abdullah's statements an "outrage "and "blatant hypocrisy." Prince Abdullah's remarks caused another failed chapter in Saudi p.r. history.

This leads to current efforts by the Saudi regime to tell Americans that they are against extremism in English - while in Arabic continuing to espouse the mantra of anti-Americanism. One of many examples that can be found on www.memritv.org is that of an interview with Saudi professor Nasser Bin Suleiman Al-Omar on a June 13 Al-Majd TV program.

Mr. al-Omar, who runs a popular Islamic Web site, www.almoslim.net, discussed the approaching collapse of America, aided by the Muslim-American community: "America is collapsing from within...Islam is advancing according to a steady plan, to the point that tens of thousands of Muslims have joined the American army and Islam is the second largest religion in America. Today, America is defeated. I have no doubt, not even for a minute, that America is on its way to destruction... It will be destroyed gradually. America will be destroyed. But we must be patient."

Another example of Saudi hatred comes from a Friday sermon at the Al-Nabawi Mosque in Medina on June 25, delivered by Sheikh Saleh Bdeir and shown on Saudi state TV: "The enemies of Islam, the Jews, Christians, [and] atheists...never cease attacking the Islamic nation...Confront your enemies' efforts with greater and stronger efforts, before these enemies become stronger."

No p.r. campaign can fix that which is contradicted by members of Saudi Arabia's royal family and government controlled press's words and deeds. To date, millions of dollars have already gone into failed attempts to improve the image of Saudi Arabia. As long as the Saudis continue to embrace an ideology of extremism, no amount of money will lead Americans to believe otherwise.

The Egyptian Press Sides With Kerry

September 1, 2004

As the November elections approach, several commentators in the Arab press have called for Senator Kerry's victory over President Bush. This trend is especially true in Egypt, which is known for its venomous anti-American press. Leading writers from Egypt's most prominent government papers such as Al-Akhbar and Al-Ahram have been particularly vocal in calling for Kerry's victory.

Professor Hasan Nafaa, who chairs the political science department at Cairo University, wrote on August 12 in Al-Ahram, "I would bet that if every eligible voter in our so-called global village had the chance to vote in the U.S. elections, the ballot would come out overwhelmingly in favor of Kerry — not out of any great fondness for him, but because of the universal loathing for his competitor. The seething antipathy for Bush that has been festering abroad has gradually seeped into the U.S. and is manifesting itself in a growing tide of opinion that is dead-set against giving that evil fanatic a second chance..."

The professor added, "...I imagine that what Bush wishes for most at this point is for bin Laden to deliver another devastating strike...The American people may not be as aware as they should be of the danger of the religious fanaticism of the American neoconservatives...His mere departure from the Oval Office will mark the beginning of the decline of the forces of extremism..."

Calling Mr. Bush a "con man," Mr. Nafaa wrote in his July 29 column: "The Bush administration may attempt to escape its predicament at home by fabricating a crisis abroad, with Iran, Syria, or both. The ideological delusions of the neoconservative clique in the White House are such that they may be tempted to push the confrontation with Iran to the brink, casting Tehran as the 'serpent's head' that must be severed in order to win the war against terrorism... Some might think that recourse to such an option is a form of madness possible only in an authoritarian government ruled by a pathological maniac...but not in a country with institutionalized checks on power and provisions for public accountability. I for one do not share this opinion."

He concluded by asking, "Can we hope that the American people will act to restrain the belligerency of this administration, at least until they have their say as to whether it stays or leaves in November? Can we hope that the American people have the sense to oust this administration in the national polls? I have no illusions..."

Deputy editor Salwa Habib of Al-Ahram explained on August 22, "If the Americans thought of the drawbacks of the [Middle East] plan, they'd consider it as another plunge into Bush's political record, and would throw him out of the White House in the coming presidential elections."

Nagi Al Shihabi, editor of the Egyptian Al-Gil newspaper, which is affiliated with the Muslim Brotherhood, was interviewed by the Iranian Al-Alam TV on June 13 about the elections, and said: "Bush declared a crusader war following the events of 9/11...The U.S. established its country over the body parts of 120 million Indians.We must first define the enemy. The no. 1 enemy of the Arab and Islamic peoples is the U.S. and not only Israel...Bush, Allah willing, will go to hell in this coming November."

The Egyptian press has used the issue of Iraq as a reason for Bush to be defeated. The editor of Al-Akhbar, Galal Dweidar, wrote on August 9, "The tragic situation in Iraq is becoming worse day by day, and the Americans are involved even more in the Iraqi swamp, and there is no way out of this catastrophe except by the departure of the rightist Bush administration, as declared by the Democrat presidential candidate John Kerry."

Prominent public figures and politicians from Egypt have also called for leadership change in America. In a recent meeting in Cairo with a visiting European diplomat, the Arab League secretary-general and former Egyptian foreign minister, Amr Moussa, was quoted as describing the liberation of Iraq as "the opening of the gates of hell," and that only a Kerry win would be able to "close those gates."

Because the Egyptian press is completely controlled by President Mubarak, its calls for Mr. Bush to be defeated and Senator Kerry to be elected should be seen as official Egyptian government policy.

Arab and Iranian TV Observe 9/11

September 8, 2004

Nawwal Nur preaches Islam in America and was interviewed about September 11 by Saudi Iqra TV on July 15. The host asked if the attacks had harmed American's image of Islam. Ms. Nur explained, "Not at all, it has not even been proven that Muslims committed it. There hasn't even been an investigation... They are confused about what happened... That is why more people converted to Islam."

As the third anniversary of 9/11 approaches, Arab and Iranian TV have been commemorating the attacks by airing such shows. Aside from the occasional denunciation of the attacks, the murder of 3,000 Americans is often observed in the region's government-controlled TV with statements by leading professors, religious leaders, government officials, and even Muslim-Americans that are often conspiracy theories stating that Arabs and Muslims were not involved and that the American government or Jews are the true culprits.

Many reasons are given for why the American government would attack its own country. Syrian researcher Tayyeb Tizini claimed on Iran's Al-Alam TV on August 16 that an intifada against globalization had broken out, and that in order to thwart it America attacked itself: "According to American and European documents, including the investigation of President Bush and his aides about 9/11, I'd like to say that 9/11 was an American action. These Americans began to understand that the new order must be marketed by a great event that would create new dangers for the world. 9/11 was for this purpose..."

Former Dean of Humanities at Egypt's Ain Shams University Mustafa Shak'a was interviewed by Iqra TV on June 16, where he attributed the 9/11 attacks to America and the Jews: "To this day, we don't know who attacked the U.S. on September 11. Why is the attack attributed to bin Laden although it has not been proven that he was involved in the operation? ... The operation was 100% American..."

At the Al-Shahid Mosque in Khartoum, Sudan on August 27, the imam Abd Al-Jalil Al-Nazir Al-Karouri discussed 9/11 during a sermon that appeared on Sudan state TV: "... When September the ninth comes [i.e. 9/11], we will again review the material we handed out about the Jewish thumbprint. Let's say, in brief, that whether the 9/9 events and the destruction of the two famous buildings in the U.S. were carried out by Israel's enemies, as the U.S. claims, or by Israeli agents, as we claim, the outcome is the same: the Jews are the cause..."

Another common conspiracy has been to state that Al-Qaeda and Osama bin Laden are puppets of America. On April 28, Lebanese Druze leader and parliamentarian Walid Jumblatt gave an interview to Al-Arabiyya TV discussing why America is responsible: "Who invented Osama bin Laden? The Americans, the CIA invented him... Osama bin Laden is like a ghost, popping up when needed. This is my opinion... Even 9/11.

In 2002, Saudi interior minister Prince Nayef blamed September 11 on "Zionists." This theory has been repeated many times in Saudi Arabia, most recently on August 16, 2004, by Saudi Cleric Sa'd bin Abdallah Al-Breik on Saudi Channel 1: "We must not inflate [the importance] of Al-Qaeda to the point of claiming that it is the main and only perpetrator of this large operation [9/11]. I'm not here to defend [Al-Qaeda], but we must not overstate this matter...It is a mistake to ignore the possibility that the Zionist hands used some people who were planted into one of the stages of this plan from this issue."

Iranian TV also frequently discusses 9/11. The Iranian station Jaam-E Jam 3 aired a program on June 15 that began with the question: "What group or organization was responsible for the events of 9/11?" The show explained how America was responsible. Similarly, on June 1, Jaam-E-Jam 1, began airing a series about September 11: "Some hours after the Twin Towers were blasted, the FBI had arrested five Israelis who had planned to blow up the New York Bridge in the Manhattan and New Jersey area. Also, the absence of 4,000 Jews [working] in the Twin Towers strengthened the claim that they took a vacation on that day..."

As each year passes following the attacks of September 11, 2001, conspiracy theories surrounding that day continue to appear on Arab and Iranian TV. While it will be commemorated as a solemn day in America as in much of the world, unfortunately, in the Middle East, it will be continued to be blamed on others.

The Power of Fatwas

September 15, 2004

Since the war on terrorism began, many fatwas have been issued by the most prominent Muslim scholars opposing American actions in Afghanistan and Iraq. According to the Encyclopedia of Islam, a fatwa is defined as "an opinion on a matter of law, the term law applying in Islam to all civil or religious matters."

Those who have issued fatwas against America include the current mufti of Egypt and leaders from Egypt's prestigious Al-Azhar. The most influential came from Muhammad Sayyed Tantawi, the sheikh of Al-Azhar, who is considered a leading authority in Sunni Islam. At a press conference he convened on April 5, 2003, Mr. Tantawi stated, "It is the right of the Iraqis to carry out any operation in defense of their homeland, whether martyrdom operations or [by] any other means."

He also encouraged volunteers from Arab and Islamic countries to go to Iraq to "support the jihad of their oppressed brethren there, because resistance to oppression is an Islamic obligation."

Shortly after an attack against American soldiers in August, Sheikh Mahmoud Ashour, a former deputy to Sheikh Tantawi, explained that Islam permits attacks against American troops: "The Western media calls these attacks 'suicide missions' but we call them 'martyrdom operations'...These so-called terrorists are defending their country from occupation. They have every right to do so. Islam gives them the right to fight the occupiers by whichever means available to them."

Most recently, it was reported on September 9 that the clerics of Al-Azhar announced that they too support the killing of American civilians in Iraq. The announcement, reported in Al-Bassa'ir, the paper of Iraq's Sunni Muslim clerics, explained that they distinguish between civilians who are not "on the scene of battle" and those who provide food, drink, and other forms of support to the military. Islam, they say, permits their killing.

It was also reported on September 2 in Al-Hayat that the influential Sheikh Yousef Al-Qaradhawi, a spiritual leader of the Muslim Brotherhood, who heads both the newly-established London-based International Council of Muslim Scholars and the European Council for Fatwa and Research, published a fatwa declaring that "it is an obligation incumbent on the Muslims to kill American citizens in Iraq, since they are in Iraq in order to assist the soldiers and the occupation forces."

Mr. Al-Qaradhawi is the most prominent Sunni Islamic leader next to Sheikh Tantawi. As news of his most recent fatwa calling on Muslims to kill Americans made its way around the Muslim world, and in particular to the terrorists fighting Americans in Iraq, the casualty number of American troops exceeded 1,000.

Ninety-three of the most prominent Islamic personalities in 30 different countries also urged Muslims throughout the world on August 23 to support "the Iraqi resistance against the American forces," and said that the purpose of the opposition is to "cleanse the house of Islam from the filth of the occupation." Among the signatories was Sheikh Al-Qaradhawi. Another influential Muslim personality, Muslim Brotherhood spiritual leader Muhammad Mahdi Akef, was quoted in the Saudi daily Al-Watan on August 20 explaining that fighting against American forces in Iraq is the realization of a religious commandment to continue the jihad.

As leading Sunni Muslim leaders continue to support the killing of American troops in Iraq, America should insist that the incitement by leading imams cease.

If this was to occur, or if the government-controlled imams in Arab states allied to America called on their followers to stop attacking American soldiers and brand such actions as haram (forbidden by Islamic law), the amount of casualties would immediately abate. This could be done in Egypt, where most of the fatwas issued against American troops originate; in Saudi Arabia, where imams frequently call for jihad against America; and in Qatar, where Sheikh Al-Qaradhawi is based.

In fact, this past week there have been some positive developments of Arab states challenging religious leaders calling for jihad against America. In Jordan, subpoenas were issued for clerics who failed to comply with a law that requires all sermons to be authorized by the government.

In Yemen, on September 11, the government announced the killing of Sheikh Hussein Badr Al-Din al-Houthi, the anti-American cleric who was the leader of the Yemeni Islamist insurgency.

Sheikh Tantawi Goes to Washington

September 22, 2004

According to the Web site of the Council on American-Islamic Relations, or CAIRs, the organization will be holding its 10th anniversary dinner October 3-4 in Washington, D.C. Invited guests include Ralph Nader, European Muslim scholar Tariq Ramadan, Grover Norquist of Americans for Tax Reform, and the head of Egypt's Al-Azhar University, Sheikh Muhammad Sayyed Tantawi.

Mr. Tantawi is arguably Sunni Islam's most prominent religious authority. He is considered a moderate representative of mainstream Islam, and has met with the pope, the archbishop of Canterbury, and the chief rabbi of Israel.

Mr. Tantawi's reputation as a moderate is due to his position of accepting the banning of the hijab in French public schools, and his sanctioning of the use of birth control. However, those who follow his comments in Arabic often find his reputation overshadowed by his anti-Semitic comments.

For example, in most of the interviews Mr. Tantawi gave following his historic meeting with the former chief rabbi of Israel, Israel Meir Lau, he referred to the rabbi and to Jews as "enemies of Islam." One such reason, he explained, is "their false claims" regarding Jewish historical connections to Jerusalem. After the meeting, Mr. Tantawi said, "I did not ask to meet with the rabbi... when he left the meeting, his face looked like his behind..."

More recently, Mr. Tantawi was quoted by the AFP on August 4, 2001, stating that the Western Wall is not really a Jewish holy site: "the temple of our Lord Solomon was not to be found underneath Al Aqsa Mosque as the Jews claim."

Islamic sources such as the Koran and Sunna are the basis for Mr. Tantawi's philosophy on Jews and Israel. In a sermon in April of 2002 he called the Jews "the enemies of Allah, the sons of apes and pigs." In an interview he gave to Al 'Ahd wa Al-Mithaq on December 12, 1998, the sheikh explained: "I've been fighting normalization for over 30 years. This is proved by the dissertation that I wrote in 1966 entitled 'The Israelites in the Koran and Suna [Following the way of the Prophet].' In this dissertation I discussed their dark history, their ways of deceiving Islam and Muslims, their atrocities as depicted by the Koran, their false claims and the ways in which they were answered by the Koran, and the punishments Allah imposed on them...There is not a single Egyptian that maintains the normalization and whoever does so is a traitor to his religion..."

Mr. Tantawi's religious rulings have also been influential regarding support for the intifada. In an interview with the UAE newspaper Al-Khaleej on October 13, 2000, Mr. Tantawi declared, "the blessed Al-Quds intifada ...demonstrated to the whole world that Allah has recruited to Al Aqsa a force that will defend it from the Jewish enemy...Permission for jihad to confront the enemy and liberate the pillaged land is an obligation upon the Muslims in every place and in every era...Jihad was legislated in Islam in order to repel the aggressor enemy...and to fight the vile. This is what our brothers are now doing in occupied Palestine. Their jihad against the enemy...and against the Israeli tyranny is a legitimate and noble jihad..."

With the most recent suicide bombing in Israel last month, which killed 19 and brought the total of Israelis killed during the intifada to over 1,200, Mr. Tantawi has also been influential in sanctioning these terrorist attacks. In April 2001, Mr. Tantawi declared, "the suicide operations are of self-defense and a kind of martyrdom, as long as the intention behind them is to kill the enemy's soldiers, and not women or children." During 2002, Mr. Tantawi called on Palestinian Arabs to intensify "martyrdom operations" against the "Zionist enemy," and described them as the highest form of jihad operations.

At this point, Mr. Tantawi upgraded his position, adding that "every martyrdom operation against any Israeli, including children, women, and teenagers, is a legitimate act according to [Islamic] religious law and Islamic commandment." In March of 2002, he further explained, "Those who blow themselves up among aggressors...and assault honor and property of Palestinians, are martyrs."

Mr. Tantawi's visit to America could provide an opportunity for American policy makers and the American Muslims who invited him, not only to ask him about the issues mentioned in this article, but also about his support for attacks on U.S. troops, about clerics under his control at Al-Azhar declaring jihad on the America, as well as his previous calls for Muslims to acquire nuclear weapons to fight "the enemy."

Campaign Under Way for Return of the Zayed Center?

September 29, 2004

On September 15 the State Department issued its annual International Religious Freedom Report, criticizing several of America's Arab allies. A paragraph of the section devoted to the United Arab Emirates noted a positive development - the closing of the Zayed Center. Indeed, for an organization to be shut down for incitement in the Arab world was an unprecedented event which should be applauded. However, immediately following the closure of the center last year, predictions arose that the center would not stay closed permanently. The spirited defense of the center following the recent State Department report may be a sign that a campaign is under way to reopen it.

From its founding in 1999 until its closure in 2003, the Zayed Center for Coordination and Follow-Up hosted events and produced studies on a variety of issues that were often anti-American and anti-Semitic. The center also hosted and worked with Presidents Clinton and Carter, and the president of France, Jacques Chirac.

Following the closure of the center, a group of Arab reformists, including Saudi journalist Sa'd ibn Salih Al-Sirhan, noted that the center got what it deserved: "The center committed professional mistakes and allowed itself to be dragged into unscholarly allegations whose only purpose was to criticize America...It embraced a number of racist lecturers...and devoted itself to provocations - provoking the Jews... and provoking America."

The controversy surrounding the center included claims it made in its publications that Americans and Jews were behind September 11. The center charged that America started the war in Iraq to coincide with the Jewish holiday Purim; the SARS virus was an American biological weapon unleashed against the world; America is ruled by Christian extremists; Jews exaggerate the number of deaths in the Holocaust; the "Protocols of the Elders of Zion" is true; the Mossad killed JFK; Zionists collaborated with the Nazis during the Holocaust, and many others.

After the facts about the Zayed Center became known, the American government pressured the namesake of the center, UAE leader Sheikh Zayed bin Sultan Al Nahyan, to close it down. On August 27, 2003, Sheikh Zayed announced that following international pressure and condemnation of its activities, the center would be closing for engaging "in a discourse that starkly contradicted the principles of interfaith tolerance."

Thousands of leading Arabs opposed the closure, calling for it to continue "being a minaret to defending Arab causes," and campaigned to keep it open. The Arab press and many Islamists blamed MEMRI. Fahmy Howedy, a prominent Egyptian Islamist journalist, said, "There is no doubting the connection between the decision to close the center and the campaign waged by the American-Zionist institute [MEMRI]..."

Kuwaiti writer Ahmad Al Dayyen also wrote in Al-Rai Al-Aam on August 19, 2003, "The name seems innocuous, but MEMRI is in fact one of the most dangerous Zionist institutions, and has considerable influence over decision-makers in the U.S. administration and in Congress. This institute is their main source of information about the situation in the Middle East...Among MEMRI's famous 'hits' in the last two years: The dismissal of Dr. Umayma Al-Jalahma, Saudi columnist for the Saudi daily Al-Riyadh, following the publication of an article which contained information from Western heritage about the Jews [i.e. the accusation that Jews use human blood for religious purposes]. Another 'hit' was the attack on Dr. Ghazi Al-Qusseibi, former Saudi ambassador to London, who published a poem in support of the Palestinian resistance [in fact, Mr. al-Qusseibi's poem supported suicide bombing] after which he was transferred from London. Recently, MEMRI scored a third 'hit,' against the Zayed Center...It launched a vicious and organized campaign against it, on the pretext that it advocates anti-American and anti-Semitic ideas. Now we are told, regretfully, that the Zayed Center will be closed. The question is, who is next on MEMRI's list of targets?"

On September 22, the Office of Information Affairs for the UAE deputy prime minister criticized the State Department report for including the closure of the Zayed Center in its annual report. The Khaleej Times reported on a statement by the deputy premier's information office, rejecting accusations against the center: " 'We would like to state unequivocally that the Zayed Center never hosted any speaker who promoted anti-Semitic views of any kind at the Center, as claimed by the 2004 report. It is unfortunate that the authors of this official U.S. government report ignored this established fact and failed to substantiate their claims with names, dates, and other evidence...' It accused the authors of the report of choosing to mimic questionable and politically motivated charges made by MEMRI...'It is high time for the U.S. administration to practice what it preaches... Frankly, if American think tanks were to be held responsible according to the same standards that were applied to the Zayed Center by the American Embassy and by MEMRI, there wouldn't be one such institution standing today on American soil.'"

This statement could mark the beginning of the re-establishment of the Zayed Center, or possibly a replacement think thank of the Arab League in its place. The London daily Al-Hayat reported on September 20, that "Egyptian academicians urged the United Arab Emirates to 'take a second look at their decision to close the Zayed Center.' The academicians, who all belong to a group they call 'The Center for Arab Research after September 11th,' announced in the office of Hosni Mubarak that it is most necessary to reopen the Center."

Within months after the Zayed Center's closure, the Arab press continued to express hopes that it would reopen. For example, the Web site Ikhwan Online published a report on December 10, 2003, about a conference of the Egyptian Journalists Union that criticized the closure of the Zayed Center. The conference was headed by Muhammad Faraj Abu Al-Nur, an author and political analyst, who criticized MEMRI for its research, which goes to "American officials and members of Congress," and led the call for the center to reopen its doors.

For a detailed list of anti-American and anti-Semitic activity of the Zayed Center, a three-part report can be found at www.memri.org.

The Intifada, 5 Years Later

October 6, 2004

Soon after the intifada was launched, leading Palestinian Arab officials began to confide to the Arabic press that it was not a spontaneous phenomenon, and that the Oslo Accords were only viewed as a temporary strategic stage, rather than an attempt at a permanent peace. The Palestinian Authority communications minister, Imad Al-Faluji, in an interview with the Lebanese daily Al-Safir on March 3, 2001, explained, "Whoever thinks that the intifada broke out because of the despised Sharon's visit to the Al-Aqsa Mosque, is wrong...This intifada was planned in advance, ever since President Arafat's return from the Camp David negotiations..."

On his way to Kuwait, where he later died of a heart attack, the former P.A. minister for Jerusalem affairs, Faysal Al-Husseini, gave what turned out to be his last interview to Al-Arabi on June 24, 2001. He explained that the Oslo Accords were a "Trojan Horse," and that the strategic goal of Palestinian Arabs "is the liberation of all historical Palestine from the River to the Sea, even if this means that the conflict will last for another thousand years."

As the intifada entered its fifth year last week much has changed since its start. The biggest loser seems to be Yasser Arafat, now holed up in his headquarters. Dozens of leading Arab writers this past year have echoed President Bush's first controversial calls for Mr. Arafat to leave, including: editor of the Kuwaiti Al-Siyassa, Ahmad Jarallah; the former editor of London-based Al-Sharq Al-Awsat, Abd Al-Rahman Al-Rashed; editor of the Egyptian Al-Akhbar, Galal Dweidar; editor of the pro-Saddam Al-Quds Al-Arabi, Abd Al-Bari Atwan, and the anti-American columnist and former editor of the London "Al-Hayat," Jihad Al-Khazen. Also, under the Bush administration, Palestinian Arab groups associated with the P.A. are listed on the State Department's terrorist list, including the Al-Aqsa Martyrs Brigades of the Fatah organization headed by Mr. Arafat.

Last week's anniversary was commemorated by many in the Arab press. On Al Jazeera, Secretary of State Powell said in published reports: "You say we blame the victims...The victims are those who are being blown up by bombs...what is the intifada in its five years of existence? Has it produced progress toward a Palestinian state? Has it defeated Israel on the battlefield?...It is time to end the intifada...the intifada has spawned terrorism and it has not achieved anything..."

The Jordanian newspaper Al-Rai published an interview with the former P.A. prime minister, Mahmoud Abbas, on September 27, 2004, in which he was quoted as saying, "Now I think that the Intifada in its entirety was a mistake and it should not have continued."

Many in the Arab world, however, call for a continuation of the intifada. Writing in Al-Ahram on September 28, Farhat Hosam Eddin expressed his encouragement: "Today marks the fifth year of Al-Aqsa intifada which...has so far presented more than 4000 martyrs...The Palestinian resistance could create fright and alarm among the Israelis through martyrdom operations in Jerusalem and Tel Aviv...The intifada has realized significant results, and made a big damage to Israel in economy, tourism, security, and politics, and raised the number of Israeli killed persons to become one against every three martyrs."

The P.A. also continues to support fighting by using incitement to violence. On September 28, 2004, the Web site of the Palestinian National Authority State Information Service stated: "The current Al Aqsa intifada was preceded by many uprisings by the Palestinian people...the resistance was born to avenge the people and the destruction that has befell the Palestinian territories.The flow of killed civilians and destroyed homes and infrastructure fuelled the Intifada and indicated that it would not stop in the near future. And today, as the intifada celebrates its fifth birthday, the Palestinian people continues to sacrifice lives and property..."

A look at the messages of official P.A. imams and main Gaza mosques also provides a glimpse into Yasser Arafat's plans for the future of his people. On September 12, 2004, Sheikh Ibrahim Mahdi appeared on P.A. TV, stating: "We must learn the lessons from [Prophet Muhammad's nocturnal journey], especially as we are fighting this bitter war with the brethren of apes and pigs, the Jews...Who will liberate the Al Aqsa Mosque once again. Al Aqsa will be liberated in the path of the Prophet Muhammad, as he said: 'The Jews will fight you and you will be imposed as rulers on them until the Jew hides behind stones or trees, and the stone or tree will say: Oh Muslim, servant of Allah, there is a Jew behind me, come and kill him except for the Gharqad, which is a tree of the Jews.'"

Arab TV Instructs on Wife-Beating

October 20, 2004

On October 5, France expelled Algerian-born Imam Abdel Qader Bouziane for telling a French magazine that Muslim husbands may beat their wives. This follows the sentencing on January 14 of the Egyptian-born Sheikh Muhammad Kamal Mustafa, the imam of the mosque of the Spanish city of Fuengirola, Costa del Sol, for publishing a book that explains that wife-beating is in accordance with Shariah law.

Sheikh Yousuf Qaradhawi, one of the most influential clerics in Sunni Islam and head of the European Council for Fatwa and Research and the International Council of Muslim Clerics, has also advocated wife-beating on multiple occasions in his 1984 book "The Lawful and the Prohibited in Islam," and on his weekly Al Jazeera program, which is popular among the surging number of European Muslims who increasingly look to religious leaders from the Middle East for religious guidance. This is particularly true with the growth of viewers watching Arab TV stations, available on satellite TV in Europe, which frequently airs shows dedicated to teaching a husband how to beat his wife. The following examples on this subject can be viewed at www.memritv.org.

Egyptian Professor Sabri Abd Al-Rauf of Al-Azhar University appeared on Iqra TV on September 13, and explained that "beating [one's wife] doesn't mean beatings with a rod or beatings that draw blood...The beatings are intended to instill fear... and declaring that he isn't satisfied with this wife."

Speaking on Syrian TV on July 26, Sheikh Abd Al-Hamid Al-Muhajir explained that the Koran stipulates when a husband can beat his wife: "The Koranic verse refers only to a disobedient wife...First you must admonish...Then comes the stage of sending her to a separate bed...If this does not help either...it is said, 'and beat them'... What's better, that she gets slapped, or that she ruins her family, herself, and society?"

Sheikh Muhammad Al-Mussayar, an Egyptian professor at Al-Azhar University, was interviewed about wife beating on Iqra TV on June 7. He described what kind of woman should be beaten: "She is a wife who rebelled against her husband's advice, and abandoning her in bed did not help."

As a member of the Egyptian Supreme Council for Islamic Affairs, Sheikh Yousuf Al-Badri explained on Egypt's Dream2 TV on September 1, "I use beatings [on my children], but symbolically. The same goes for women: 'Admonish them and send them to beds apart and beat them...' There are beatings in the Koran and in the Sunna...This means we're allowed to beat."

On August 26, Qatar TV aired a panel discussion that included Dr. Ibrahim Elias, and the Director of The Women's Development Society, Imam Bibars, who discussed a study she performed in the Arab world: "I'd like to say that I found something that took me by surprise. I call it 'a culture of the electricity cable.' The men in the study did not know one another, but they all used to beat their wives with electricity cables. These cables are large and they would beat their wives." In defense of beatings, Dr. Elias, a lawyer, explained, "If you beat your wife and it's only light beatings in order to set things straight - that's it...We tell him, 'They are not considered an assault, but discipline.'"

Responding to the question, "What do you mean by light beatings?" he gave an example of when a man should be beat his wife: "For example, a man comes home from work and finds his wife watching TV. She doesn't even get up to make him food. He tells her once, twice, and asks again. If only once he would raise his voice and beat her, she would get up to prepare food for him and by the next day she'd be obedient. This will last for a week and when she forgets, he will remind her."

The following day on the same channel, a religious leader detailed three types of women who deserve beatings: "The first type is a girl who was brought up this way...The second type is a woman who is condescending towards her husband... With her, too, only a rod will help. The third type is a twisted woman who will not obey her husband unless he oppresses her, beats her, uses force against her, and overpowers her."

Unfortunately, the examples mentioned in this article are the rule, not the exception. TV shows dedicated to husbands beating their wives can be viewed regularly on Arab TV.

Arabs and Muslims for Kerry

October 27, 2004

Over the past few months this column has discussed how the Arab and Iranian press have been following the upcoming presidential campaign. One common theme in the press has been to explicitly call on Arab and Muslim voters in America to vote for Senator Kerry. For example, in Egypt's Al-Ahram Weekly on April 18, Arab-Israeli Member of Knesset Azmi Bishara called on Arab-Americans to "rethink their alliance with Bush" and "shift their votes from Bush to Kerry."

More recently, on October 15, the former prime minister of Malaysia, Mahathir Mohamad, who last year stated that "Jews rule the world by proxy," called on Muslims in America to support Mr. Kerry: "Vote Bush out of office. It is truly a [religious worship] that you would be performing..." He added that there is a duty to ensure that "Bush will not be able to determine our fate for four more years...I would like to appeal to you, the Muslim citizens and voters of America, to be united and to cast your votes against Bush, in the name of justice, in the name of Islam."

In an article titled "U.S. Muslims Urged to Back Kerry," Al Jazeera reported on October 24 how American Muslims are being encouraged to post a protest vote for Mr. Kerry, even though they do not particularly like him. The American Muslim Taskforce on Civil Rights and Elections was cited as one group calling on its members to vote against President Bush. This organization is an umbrella group, whose members include The American Muslim Alliance, or AMA, Council on American-Islamic Relations, or CAIR, Islamic Society of North America, or ISNA, and other groups.

Many Arabs and Muslims living in America and writing in the Arab press are advised how to vote. Writing in Lebanon's Daily Star on August 20, Muqtedar Khan, a fellow at the Brookings Institute's Saban Center for Middle East Policy, called for Mr. Bush to leave office: "... What September 11 has done is give the Bush administration a blank check to act recklessly while responding to the fears and opportunities that presented themselves... Unfortunately the policies of the Bush administration have projected it as a major threat to international law and respect for human right." Mr. Khan concluded by stating, "Bush's policies have made America uglier without making it safer. Thank you, Mr. President, but no thank you. It's time for you to ride into the sunset, cowboy."

James Zogby, president of the Arab American Institute, writing in the Saudi daily Arab News on October 20th in a column titled "Why I'll Vote for Kerry," stated: "This November, I will vote for John Kerry for president of the United States...A dangerous and arrogant ideological group in the White House and the Pentagon lied our nation into war...The American people, the Iraqi people, the world, and our nation's image and standing in the world, are paying the price for this deceit and arrogance...I believe that Arab-Americans are better served by the... Democratic Party than they are by the Republican coalition that is driven by the religious right and neo-conservative ideologues."

An article in the Saudi Gazette on October 12 by Islam Abdullah, director of the Muslim Electorates Council of America and of the Islamic Society in Nevada, quoted Muslim-American voters such as Captain Shehzad Wahid of California, who similarly explained his disdain for "neo-conservatives" and members of the "Christian right," stating, "Kerry is better than Bush. I will vote for him."The writer explained that his "sentiments were expressed by Muslims all over the U.S.," who are particularly concerned that a second Bush term might produce "Supreme Court judges...representing neo-cons and [the] Christian right."

In an article titled "Why I am Supporting John Kerry," Islam Siddiqui, who served in the Clinton administration as an undersecretary in the Department of Agriculture, wrote in Arab News on October 19: "It is time to end the 'Era of John Ashcroft,' and restore our civil liberties...It is clear that we cannot live under another four years of the Bush-Cheney-Ashcroft administration. John Kerry and John Edwards recognize that American Muslims are a vital part of the American mosaic...American Muslims and Arab-Americans cannot stand by the sidelines and allow an administration in the White House that will continue to trample on our civil liberties..."

As Election Day approaches, polls by Arab and Muslim American organizations suggest that Mr. Kerry will receive the majority of their community's votes. As one Muslim American political activist was quoted by Al Jazeera: "A lot of Muslims want to vote on principle for Nader...Most of the community is not excited about voting for Kerry...Unfortunately, we have no choice."

A Response to CAIR's Findings on Anti-Muslim Sentiment

November 3, 2004

The Council on American Islamic Relations released a poll in October, concluding that nearly one-third of Americans hold negative stereotypes of Muslims, such as: Islam encourages oppression of women; Islam teaches violence and hatred; Muslims value life less than other people, and Muslims teach their children to hate unbelievers.

Why might Americans hold such beliefs?

The charge that Islam encourages oppression of women: In the Middle East, the beating of women is sanctioned regularly on Arab TV shows (see www.memritv.org); polygamy is legal and women are often forced into marriages; in Saudi Arabia, women have almost no rights and cannot go out in public without a male relative, or even drive a car; in Jordan, honor killings are sanctioned by law; in Egypt, many young girls are forced to undergo female circumcision (genital mutilation), and yesterday, Dutch filmmaker Theo van Gogh (great grandson of Vincent van Gogh) was murdered, allegedly by a man with Moroccan-Dutch citizenship, following a documentary Van Gogh made about the treatment of Muslim women.

The charge that the Muslim religion teaches violence and hatred: The world's three leading mainstream Sunni Muslim figures all espouse violence and hatred.

Sheikh Abd Al-Rahman Al-Sudayyis is the Saudi government-appointed imam of the Grand Mosque in Mecca. Themes of his sermons are characterized by confrontation toward non-Muslims. Mr. Al-Sudayyis calls Jews "scum of the earth" and "monkeys and pigs" who should be "annihilated." Other enemies of Islam, he says, are "worshippers of the cross" and "idol-worshipping Hindus."

Sheikh Muhammad Sayyed Tantawi is head of Egypt's Al-Azhar University, Sunni Islam's most prestigious establishment. Under Mr. Tantawi's leadership, Al-Azhar has declared jihad against America repeatedly, and has called for suicide attacks against American forces in Iraq as recently as August. He has called Jews "apes and pigs." CAIR invited Mr. Tantawi to speak in Washington, D.C., at its 10th annual dinner last month.

Sheikh Yousef Al-Qaradhawi is considered the spiritual leader of the Muslim Brotherhood movement, and increasingly influences European Muslims through his roles as head of both the European Council for Fatwa and Research and the International Council of Muslim Clerics. He has called for Islam to conquer Europe, expressed support for suicide attacks, and most recently issued a fatwa calling on Muslims to attack American troops.

The charge that Muslims value death more than life: A Palestinian textbook published during the Oslo accords teaches second-graders, "As much as you love life, we love death - and the desire for it." Leading Muslim clerics often refer to the love of death. Chief Palestinian Authority cleric, Sheikh Ikrimeh Sabri, who is the spiritual leader of Islam's third holiest mosque, explained in 2001, "As much as you love life, the Muslim loves death and martyrdom."

The charge that Muslims teach their children to hate unbelievers: A look at education systems in the Arab world reveals widespread incitement to hatred against non-Muslims. Examples include a Saudi textbook for eighth-grade students quoting the Koran that Jews and Christians were cursed by Allah and turned into apes and pigs. Another Saudi textbook for fifth-graders instructs the students that the only true religion is Islam, adding, "The whole world should convert to Islam and leave its false religions, lest their fate will be hell." Even though Saudi Arabia claims it no longer teaches hatred in its schools, this is not true.

Responding to the poll, CAIR doesn't address why Americans might have legitimate concerns about some aspects of the Muslim world, such as ones listed in this column. Instead, it placed the onus on Americans to learn about Islam - which is a code word for the Islamic concept of da'wa, converting non-Muslims.

CAIR's executive director, Nihad Awad, explained one positive development about the poll results: "The majority of Americans are open-minded and receptive" to learning more about Islam, including CAIR's efforts to teach Islam to Americans. It should be noted that in an interview on November 16, 2001, Mr. Awad triumphantly told the Saudi newspaper Ukaz that "34,000 Americans have converted to Islam following the events of September 11, 2001." Alaa Bayumi, Director of Arab Affairs for CAIR, also wrote in Al-Hayat the same week, adding "The Americans are showing increasing willingness to convert to Islam since September 11..."

Americans are correct to oppose objectionable practices and ideologies in the Arab and Muslim world that lead to terrorism and hatred. It is unfortunate that CAIR has failed to do so. This, of course, does not justify prejudices against individual Muslims.

Ramadan Incitement in Arab Press

November 10, 2004

During the holy month of Ramadan this year, some interesting developments have occurred in the Arab world. In Saudi Arabia, the Interior Ministry headed by Prince Nayef decreed on October 12 that all non-Muslim residents must follow the Muslim law of fasting during Ramadan. Anyone who fails to do so will be deported.

In Egypt it is customary for vendors at markets to offer special brand names for the popular dates that Muslims commonly eat to break the daily fast. Dates are usually named after prominent world personalities, both popular and hated, as well as other subjects that made news during the year.

Arab newspapers such as Lebanon's Daily Star, Qatar's the Peninsula, Egypt's Al-Akhbar, and the Web site Islam Online are reporting that the most expensive dates this year are named after Osama bin Laden and Abu Musab al-Zarqawi. Other popular brands this year are "Saddam Hussein," "Iraqi Resistance," "Palestinian Intifada," and "Al Aqsa Mosque." Last year's "Saddam Hussein's Weapons of Mass Destruction" and "George W. Bush" are currently off the market, as is last year's most popular "Jacque Chirac," which was removed reportedly due to his stance regarding the hijab in France as well as France's insistence that Syria withdraw from Lebanon. "Yasser Arafat" dates are also absent this year, while for the third year in a row "Sharon the Butcher" dates are the cheapest, worst quality, and most damaged. Another of the "bargain section" this year are named after John Kerry and are reportedly "the longest," to resemble his chin.

The month of Ramadan is the busiest TV-viewing time of the year in the Arab and Muslim world. TV stations air religious programs depicting Islamic history and battles to inspire today's Muslims to jihad. One example includes TV host Hisham Khreisat, who had a monologue on the Saudi Al-Majd TV channel on October 17, 2004: "This great month is a month of fasting, a month of blessings and victory, a month of jihad and martyrdom...They have always been battles of victory and liberation. The most decisive battles in the life of the nation were always in Ramadan. Ramadan...is a month of heroism, work, effort, and sacrifice, a month of martyrdom and martyrs. Thus we have been taught by history...[that we] will remain alive thanks to the blood of [our] martyrs..."

Also shown on TV are annual miniseries devoted to an assortment of subjects, which are often anti-Semitic. "Al-Shatat," a 29-part Hezbollah-produced series, made with the cooperation of the Syrian government, aired last year on Hezbollah's Al-Manar TV. The program depicts a centuries-old plot by Jews to take over the world, and was based on "The Protocols of the Elders of Zion." It included scenes such as a rabbi killing a Christian boy to use his blood to make matzoh. "Knight Without a Horse," a 41-part series also based on the "Protocols," was produced by an Egyptian, Muhammad Subhi, and broadcast during Ramadan of 2002 by Egyptian state TV, the Egyptian "Dream" channel, Iraqi state TV, and Hezbollah's Al-Manar. Both of these programs are being aired in Persian this Ramadan on Iranian TV.

This year Arab TV stations are also broadcasting a series on the life of the Palestinian Arab Hamas terrorist Yahya Ayyash, nicknamed "The Engineer" for his skill at creating bombs. Mr. Ayyash built the bombs used in numerous suicide bombings from 1994-96 that killed more than 100 Israelis.

The TV series making the most news this Ramadan is the Jordanian series "Road to Kabul," which tells the narrative of modern Afghanistan from Soviet occupation through the American invasion in 2001. Islamist groups who considered the series critical of their actions opposed its airing, and on one Islamist Web site a group called "The Brigades of the Mujahideen in Iraq and Syria" published a threat: "We would like to strongly warn everyone who contributed to the production of the series, including actors, producer, and photographers, that the series should not contain anything that slanders the honorable phenomenon of Taliban..." Shortly after this statement was released, TV stations pulled the series from their airwaves.

The removal of the program comes in contrast to the response to Western demands to cease airing previous series. In 2002, the State Department called on the Egyptian government to prevent the broadcast of "Knight Without a Horse" - a demand that was rejected by the Egyptian information minister, Safwat Al-Sharif. As this year's Ramadan reaches its peak, the Arab press has continued its trend of using the holiday for incitement to jihad and propagating anti-Semitism. It is doubtful that that the Arab press will make any significant changes by next Ramadan.

Yasser Arafat's Martyrdom Legacy

November 18, 2004

"They will not take me captive or prisoner, or expel me, but [I will go] as a martyr, martyr, martyr. O Allah, give me martyrdom." So said Yasser Arafat in an interview in 2002, and many other times.

While the last week has witnessed much talk over Yasser Arafat's legacy, clearly he will be remembered for introducing both suicide bombers and child martyrs as political tools. Arafat was able to do so by creating a culture of incitement, explicitly calling on Palestinian Arab children to become martyrs through indoctrination in textbooks, sermons, and the press, all under his control.

There have been more than 30 suicide bombers younger than 18-yearsold, some sent by Arafat's Fatah Al Aqsa Martyrs Brigade, and numerous children younger than 13 who have been apprehended during suicide attack missions. The most recent child suicide bomber was a 16-year-old who killed three in Tel Aviv on November 1.

Since Arafat entered the West Bank and Gaza in 1994, he often spoke directly to Palestinian Arab children, calling them his "lion cubs" and "flowers" and telling them to become martyrs. An example of Arafat's incitement of children includes the speech he gave to a group of children that aired on Palestinian Authority TV to mark the International Day of the Child on July 1, 2003: "Oh my dears, oh lion cubs of Palestine, oh, flowers of Palestine, oh, future of Palestine, oh power and splendor of Palestine. ...One Shahid from those [fighting for Jerusalem] is equal in value to 40 Shahids...This is the privilege Allah has given to you, oh boys and girls of Palestine..."

During a speech about the importance of child martyrs on December 18, 2001, which was reported by WAFA, Arafat stated: "Do you know what a mother of a martyr does when she is informed of the martyrdom of her son? She goes out to the street with cheers of joy, saying: 'Allah be praised, my son, that you married Palestine rather than your cousin.' This is the Palestinian people."

Pointing to a young girl, Arafat stated, "This little cutie is the first of our soldiers...One of our cubs or one of our flowers will wave the flag of Palestine, Allah willing, over the walls of Jerusalem, its churches and its mosques, whether someone likes it or not, and whomever does not like it can drink the water of the Dead Sea...

[Islamic tradition states that] 'One martyr of these is worth seventy' [on any other front of jihad]..."

In a performance by Palestinian Arab schoolgirls for Arafat on May 19, 1997, a young girl sang to Arafat: "I finished practicing on the submachine gun of Return, and I trained my friends from among the children and the youth, and we swore to take vengeful blood from our enemies..."

More recently, Palestinian Authority TV aired a special program on October 10, 2003, of an Arafat speech from the muqata to Palestinian Arab children attending summer camps: "Here it is. Here it is, the Shabiba [youth] of Fatah. With soul and blood we will redeem you, oh Palestine. Millions of martyrs marching to Jerusalem..."

Arafat's position regarding inciting Palestinian Arab children to become suicide bombers was accepted within Palestinian Arab culture because he always stated that he, too, wanted martyrdom. One example is a speech aired on Al Jazeera on March 29, 2002, following the Palestinian attack on a Passover seder, which killed 30 people: "We ask Allah to grant us martyrdom, to grant us martyrdom. To Jerusalem we march, martyrs by the millions." Arafat added: "[If] they decided to take me as a prisoner, a deportee, or to kill me - no. I say to them [that I will be] a martyr, a martyr, a martyr...One of their martyrs [who falls in the battle for Jerusalem] is worth 40 martyrs. ...Allah, give me martyrdom. ...This is the path I have chosen..."

In an interview shown on Egyptian TV and PA TV that day, Arafat explained: "We are all seekers of martyrdom. ...As I told you: To Jerusalem we march, martyrs by the millions. ...Man, don't wish me safety! Pray for me to attain martyrdom! Is there anything better than being martyred on this holy land? We are all seekers of martyrdom..."

In an interview with the Arabic-language daily Al-Hayat on October 11, 2002, Arafat was asked if he was worried about being expelled by Israel. He answered: "... Have you forgotten my motto? They will not take me captive or prisoner, or expel [me], but as a martyr, martyr, martyr. O Allah, give me martyrdom."

Despite the fact that Arafat didn't martyr himself as he had promised, his message for Palestinian Arab children to become martyrs will most likely live on as an inspiration for future generations.

Palestinians Talking Peace?

November 24, 2004

Yasser Arafat's death has created an unprecedented amount of optimism in the West regarding the establishment of a Palestinian state and the possibility of peace. Yet amongst Palestinian Arab officials there is little talk of peace, as numerous officials have endorsed the continuation of Arafat's "jihad" against the Jewish state.

Some members of the Palestinian Arab establishment close to Arafat are now stating in public that he never really wanted peace, and instead used the peace process as a strategy to destroy Israel in phases. The Palestinian ambassador in Iran, Salah Al-Zawawi, explained in an interview on Iranian Al-Alam TV on November 12 that Arafat "knew that this path is the path of martyrdom and jihad. He knew that this great cause requires martyrs, not leaders...He fought the jihad and we saw him in many battles...If you ask me what will surely be the end of this Zionist entity, I will say to you that this entity will disappear one of these days... It's a matter of time...Our phased plan, which I already mentioned, is to establish an independent sovereign Palestinian state with Jerusalem as its capital..."

Similarly, Palestinian Arab analyst Yunis Udeh told London's ANB TV November 11: "When we told him that the road to Oslo would mean the termination of the Palestinian cause, he said, 'I am hammering the first nail in the Zionist coffin.'...I asked him how. He said: 'I will go to Gaza, I will return to Palestine...'"

Fatah Supreme Council Member Abu Ali Shahin also hinted in an interview on November 13 on Lebanon's Al-Manar TV that Arafat considered the Oslo Accords a strategic move to destroy Israel: "Yasser Arafat led a revolution, a revolution of a barrel of gun powder alongside a barrel of petrol...But when Yasser Arafat saw that the U.S.S.R....collapsed without a single shot being fired...Arafat understood this great international game. He made a 180-degree turn...He accepted...Madrid, and after it Oslo..."

Countless Palestinian Arab officials have also spoken about continuing violence against Israel. Fatah leader Hussein Al-Sheikh spoke to Al-Arabiyya TV on November 11: "The gun Yasser Arafat raised...will be raised by...the Palestinian people, so they continue to believe that the gun is the way to get rid of this occupation, the shortest way to get rid of this occupation. This is Abu Ammar's promise and this is his will, and we will continue to be true to them."

Also on November 11, Al Aqsa Martyrs Brigade leader Raid Al-Aidi said on Al-Arabiyya TV, "We call from here to all the heroes...strike this occupier anywhere, with no holds barred. We...will direct our painful blows against this monstrous entity. The Palestinian state will be achieved only by strengthening the resistance... This occupier understands only the language of gunfire and gun powder and we will teach this occupier, Allah willing, a lesson as we have taught it in the past, in Tel Aviv, Hadera, and everywhere. We will escalate our blows against this occupier..."

In the same program, Fatah Central Committee member Hani Al-Hassan explained that, "In Fatah we have a rule: The armed struggle sows and the political struggle reaps...Therefore, when Oslo didn't bring results, the sowing came in the form of the Intifada...We will see now whether the political situation allows us to reach political results and to bring about a change in our favor. Otherwise, we will go back to sowing."

Quoting former Egyptian president Abd-Al Nasser, "what was taken by force will be restored only by force" is how the new leader of Fatah, Faruq Al-Qaddumi, described the Palestinian Arab strategy against Israel on Al-Arabiyya TV on November 14. Mr. al-Qaddumi has considerable popularity on the Palestinian street for never accepting Oslo. With his naming as leader of Fatah, Mr. al-Qaddumi is openly challenging Mahmoud Abbas and Ahmad Qurei to be Arafat's successor. As he stated, "Anyone who thinks that I have abdicated my authority is mistaken."

He explained Fatah's position about Hamas: "The Hamas movement is our friend. It is a...movement of heroes. It is part of the national Palestinian movement. No...Fatah member could possibly harm Hamas." Mr. al-Qaddumi is also close to Hezbollah, and during a meeting with Sheikh Nasrallah on September 4, 2003, they discussed "a cohesion between the Lebanese and Palestinian resistance." He has also called for an attack on American interests throughout the world. U.S. officials intimately involved in the Oslo Accords now publicly state that more attention should have been paid to the issue of Palestinian Arab incitement, and what the Arabs were saying amongst themselves in Arabic. With Arafat gone, we should be paying close attention to his heirs to understand their true intentions.

Egyptian Support for Killing American Soldiers in Iraq

December 2, 2004

Support for the Iraqi "resistance" against American troops has become the cause celebre in the Arab press. Such support has not only come from Baathist and Islamist sources, but unfortunately also from our Arab allies such as Egypt, from government and opposition spokesmen. One notable example is from the most prominent Egyptian government newspaper Al-Akhbar, which reported last year that American troops were cannibalizing Iraqi corpses.

A more recent example is Egyptian Labor Party head Magdi Ahmad Hussein, who appeared on Al Jazeera TV on October 17. He cited the Prophet Muhammad, the spiritual founder of Wahhabi Islam Ibn Taymiyya, and the prominent 20th century Al-Azhar Sheikh Shaltout, for setting a precedent for attacks against American troops: "The violence is currently directed at the occupation. This is legitimate violence. This is jihad against occupiers...Both the Koran and the prophet's biography permit the killing of prisoners. This exists in our Islamic law...50 years ago, even before the American army arrived in Iraq, Sheikh Shaltout said, 'Anyone working in the enemies' military camps and factories is one of them. He's an enemy and he may be killed...' The mujahed should be there, and the cleric should be there, like Ibn Taymiyya, who set out with the mujahedeen to the front lines..."

The current sheikh of Al-Azhar, Muhammad Sayyed Tantawi, spoke on Lebanese New TV on August 19 to support killing American troops: "Anyone who blows himself up amongst the enemy that wants to kill him and he can find no means of defending himself except blowing himself up amongst these soldiers who are occupying his land and destroying his home, and so he blows himself up amongst these aggressive soldiers, is a shahid, shahid, shahid..." When asked specifically about U.S. troops in Iraq, he avoided giving a direct answer, instead saying, "Don't specify countries and names. I determined a general rule which is based on religious law and that applies to everybody."

An international relations expert from Egypt's most prominent thinktank, the

Al-Ahram Center, Said Al-Lawindi, appeared on Egypt's Channel One on November 16, to defend those fighting American troops: "This makes it [a] necessity to distinguish between men of resistance...and the terrorist[s] who leave only scorched earth behind them. I maintain that this is the logic of the American cowboy...The U.S. is a rogue state. It accuses others of [being] rogues, while in fact the U.S. is the greatest rogue state in the entire world."

The deputy leader of the Muslim Brotherhood in Egypt, Mahmoud Al-Sayyid Ahmad Al-Habib, gave an interview to Hezbollah's Al-Manar TV on April 8 praising suicide bombings against American troops: "Martyrdom operations...by the Iraqi resistance - these redeem self-confidence and hope, because a nation that does not excel at the industry of death does not deserve life." Abd Al-Muni'm Abu Al-Futuh, from the General Guidance Council of the Muslim Brotherhood, appearedon Al Jazeera TV on April 25, adding, "Regarding the resistance, I say clearly and without evasion that we support armed resistance...in Iraq against the American occupier, 100%...The Muslim Brotherhood in Egypt and around the world wish to carry weapons against the American occupation in Iraq..."

During fighting this summer in Najaf, Egypt's Muslim Brotherhood issued a public appeal to support the insurgents fighting American soldiers. The statement expresses solidarity "with the noble and brave national Islamic resistance, and call on them to join ranks in the struggle against the occupation." The leading figure among the signatories is Sheikh Muhammad Mahdi 'Akef, leader of the Muslim Brotherhood. The appeal was signed by leading Islamist clerics throughout the Arab world.

Writing in Al-Ahram on November 25, Anis Mansour criticized Arab press outlets that state that American soldiers should not be killed. He explained, "Spoiled American soldiers, who have traveled thousand of miles and [are] stationed in desert areas, believe they should not be killed while protecting oil fields and lands as if Iraq is an American property...'Crusade' is the exact term to describe the current U.S. race against time [to] destroy Arab unity and let them follow the example of former Soviet Union when [it] collapsed."

The Egyptian government has done almost nothing to stop incitement for attacks against U.S. troops. Despite being America's main Arab ally and the recipient of nearly $2 billion annually in American foreign aid, Egypt is acting more like an opponent than a friend of America.

Rights of Jews To Jerusalem Are Denied

December 8, 2004

Within the Arab world, there has been an ongoing effort to dismiss any Jewish connection to Jerusalem. Leading figures from Palestinian academic, religious, and political establishments have been particularly outspoken on this subject.

Speaking on Palestinian Authority TV on November 21, Palestinian Arab historian Issam Sisalem, one of the two "home historians" on PA TV, compared Jews to "parasitical worms" who make "false claims" and tell "Biblical lies." Regarding Jerusalem, he explained that it has always belonged to Arabs, adding, "Everything on our land is holy to us and we will defend it and never be deceived, nor will we argue about biblical lies that penetrated the minds of some orientalists. Jerusalem has been the capital of our capitals since the dawn of history, and will remain so. We won't share it...The Torah disparages all of the prophets. It presents David as an adulterer and Solomon, too. They say that the temple was here. What temple and what shtemple?! What archeological remains?"

The theme that there is no Jewish connection to Jerusalem is also common in sermons and statements by Palestinian Arab religious figures. For example, Sheikh Ikrima Sabri, the mufti of Jerusalem and the highest-ranking Palestinian Arab religious figure gave an interview to the German Die Welt on January 17, 2001, stating, "There is not the smallest indication of the existence of a Jewish temple on this place in the past. In the whole city, there is not even a single stone indicating Jewish history...The Jews do not even know exactly where their temple stood." When the interviewer asked why he could not respect the Jewish connection to the place, he added, "It is the art of the Jews to deceive the world...There is not a single stone in the Wailing Wall relating to Jewish history. The Jews cannot legitimately claim this wall, neither religiously nor historically."

More recently speaking to Egyptian Dream TV on May 9 he added, "The Al-Buraq Wall [the Western Wall] is part of Al Aqsa Mosque. It is part of a wall, the wall [surrounding] the mosque. Just like any house with a wall, this wall belongs to the house. The Al-Buraq Wall, is part of the Al Aqsa Mosque...The Jews pray in front of the wall, on the outer side of it...but the wall belongs to the Muslims."

The Palestinian denial of the Jewish affiliation to Jerusalem is also common among political leaders. Yasser Arafat frequently claimed that the Jews made up a religious connection to both Jerusalem and the Wailing Wall. One example includes an interview he gave to the London Arabic daily Al-Hayat on October 5, 2002. He explained, "They found not a single stone proving that the Temple of Solomon was there, because historically the Temple was not in Palestine."

Mahmoud Abbas, Arafat's successor who is now chairman of the Palestine Liberation Organization's executive committee and is running for the Palestinian presidency, supports Arafat's theories on Jerusalem. Speaking to the Israeli-Arab weekly, Kul Al Arab on August 25, 2000, he stated, "Anyone who wants to forget the past cannot come and claim that the temple is situated beneath the Haram. They demand that we forget what happened 50 years ago to the refugees - and I speak as a living, breathing refugee - while at the same time they claim that 2,000 years ago they had a temple. I challenge the claim that this is so."

Representing Palestinian academia, a researcher and editor-in-chief of the Al Quds Al-Sharif Encyclopedia, Hassan Ali Khater, spoke about the Wailing Wall at the now defunct Zayed Center on January 27, 2002: "Israelis are falsifying history by inscribing Jewish inscriptions on rocks then calling international experts to rediscover [them] as Jewish monuments."

Another report by the Zayed Center, titled "Al-Buraq Wall not Wailing Wall," stated, "This study is [to] refute the false Zionist allegations in regard to Palestine...and shows the falsehood of the Zionist religious claims and anthropological fabrications. The study ...underlines that Al Aqsa Mosque was built more than a thousand years before Solomon, giving evidence that refutes the Zionist allegations that the mosque was constructed on the ruins of Solomon's Temple...all of which demonstrate that the Zionist alleged rights to the wall are simply baseless."

The issue of Jerusalem can be expected to remain at the center of the Arab-Israeli conflict. The city's history is recorded in religious, archaeological, and scientific research. No solution to the complex issue of Jerusalem will be found by evading such research.

Lebanese MP Treated Like King in Paris

December 16, 2004

Over the past few years, France has won the reputation as being the leading anti-American voice of the Western world. Lately, news has been trickling out of France, suggesting that President Chirac has taken his opposition to America to a new level. One example includes allowing the pro-Baath pro-Saddam group "La Resistance" to operate out of Paris. The group's main goal is to support killing American troops in Iraq.

Another example occurred less than two weeks ago when the French government hosted Lebanese parliamentarian Walid Jumblatt, Head of the Progressive Socialist Party, who enjoyed an unprecedented state visit to France. As Ahmad Al-Jarallah, editor-in-chief of the Kuwaiti daily Al-Siyassah, reported on December 4, "Walid Jumblatt...[was] recently given a send-off befitting a state leader at the Elysee Palace by French President Jacques Chirac." According to the Lebanese Daily Star, "Jumblatt's visit to the French capital was considered by some political figures as historic, and as bringing him merit on the regional and international levels."

This is significant because Mr. Jumblatt is known for his vehement anti-American statements and antagonistic attitude toward America. On November, 19, 2003, it was reported that the state department cancelled Mr. Jumblatt's diplomatic visa following revelations that he expressed regret that the deputy secretary of defense, Paul Wolfowitz, was not killed in a missile attack during a visit to Baghdad.

More recently, Mr. Jumblatt gave an interview to Al-Sharq Al-Awsat on February 12, 2004, in which he said: "We are all happy when U.S. soldiers are killed [in Iraq] week in and week out. The killing of U.S. soldiers in Iraq is legitimate and obligatory." The Progressive Socialist Party leader has also said he felt "great joy" at the 2002 destruction of the U.S. Space Shuttle Columbia, because it carried an Israeli astronaut.

The Lebanese MP is also known for espousing conspiracy theories against America. On April 28, 2004, he gave an interview to Al Arabiyya TV, in which he detailed how America was really behind September 11: "Who invented Osama bin Laden?! The Americans, the CIA invented him so they could fight the Soviets in Afghanistan together with some of the Arab regimes. Osama bin Laden is like a ghost, popping up when needed. This is my opinion."

Mr. Jumblatt was asked "Even 9/11?" and answered: "Even 9/11...Why didn't the sirens go off when the four hijacked planes took off?...The U.S. always needs an enemy...According to this plan or ideology of the born-again Christians who formed an alliance with Zionism - Islam is the monster, Islam is the target."

In addition to hating America, Mr. Jumblatt has also spoke against the countries that support America. Lebanon's Daily Star published a February 3, 2003, article quoting him as saying that the true axis of evil is one of "oil and Jews," calling President Bush a "mad emperor," and insulting Prime Minister Blair, Prime Minister Berlusconi, and Jose Maria Aznar, the former Spanish prime minister: "The oil axis is present in most of the U.S. administration, beginning with its president, vice-president, and top advisers, including [Condoleezza] Rice, who is oil-colored, while the axis of Jews is present with Paul Wolfowitz."

In the interview, Mr. Jumblatt described President Bush as someone who "considers himself God's deputy on Earth, threatening and classifying the world [into different camps], and relying on his imperial power...How dangerous emperors are when they go mad... In the same axis we have the trustworthy servant, the imperial servant...pleased with himself and his idiotic laugh, his peacock appearance, none other than Tony Blair...Also joining this axis is the comprador Mussolini of the 21st century, the prime minister of Italy today, Silvio Berlusconi, who seems to want to renew the empire of the Caesars...To complete the picture, we have Spanish Prime Minister Jose Maria Aznar, the Spanish neo-rightist...Aznar and Blair spend a lot of time in front of the mirror every morning, it seems, so that their hair is parted perfectly...People who pay that much attention to their appearance are fascists by nature. Or they have psychological or sexual complexes."

The French government's providing a hero's welcome to Mr. Jumblatt, a self-proclaimed enemy of America, represents something far beyond a simple diplomatic disagreement.

Muslim Conspiracy Theories Are Running Rampant

December 23, 2004

The following article is the fourth in a series highlighting recent conspiracy theories from the Middle East. Following up on the previous reports from the summer and fall, conspiracy theories have abounded, specifically, America is to blame for the September 11 attacks; "Zionists" spread AIDS; Abu Musab Al-Zarwaqi is an American propaganda invention; the CIA is writing sermons for Egyptian imams and the Jews are rewriting the Koran, and more recently, Israel killed Yasser Arafat.

Conspiracy theories surrounding Iraq have been especially popular - in fact, MEMRI has been the subject of several. For example, Hezbollah's Al-Manar TV broadcast a program about MEMRI's alleged activities in Iraq, claiming that MEMRI is commanding "Zionist squads specializing in car bombs." Iranian TV Channel 1 claimed that MEMRI was responsible for the bombing of Iraqi oil fields.

One of the most popular conspiracy theories about Iraq deals with claims of Jewish/Israeli/Zionist involvement in terror attacks and fighting Iraqis. For example, in a sermon broadcast on Sudan TV on September 19, Sheikh Abd Al-Jalil Al-Karouri preached, "There are currently 1,000 Jewish soldiers in Fallujah. Among them are 37 rabbis. The rabbis raise the morale among the Jewish soldiers fighting in Fallujah. So America is fighting for Israel...The Jews came to conduct urban warfare on behalf of America, and America is supporting them in exchange."

Iran's Mehr News Agency published an article October 5, explaining that "Israeli commandos played an active role in the occupying forces' attacks on Najaf, Sadr City...Fallujah, and some areas in Tikrit, Ramadi, and Baqubah." The article added, "Israelis have taken over some military operations...According to certain reports, there are over 200 Zionist troops in Iraq dressed in U.S. Army uniforms."

Similarly, Iran's Al-Alam TV broadcast an interview with Jordanian MP Adnan Hasouna on October 28, in which he claimed that the Mossad is responsible for attacks in Iraq: "The Israeli Mossad, the Jewish Mossad, and others strive to distort the image of Islam by striking the National Guard soldiers and civilians."

The December 2 edition of the Egyptian daily Al-Ahram reported that Mohamed Selim El-Awwa, a prominent Egyptian Islamist and secretary-general of the International Association of Muslim Scholars (IAMS), claimed that the Iraqi resistance had been infiltrated by "Zionist and international intelligence services." Mr. El-Awwa explained that Muslims would not decapitate a hostage. "It is no coincidence that decapitations or the brutal killing of innocent hostages always occur just after a scandal exposing the violations of the occupation forces."

In the Saudi daily Arab News on September 7, Hassan Tahsin wrote an article titled, "Danger Of Israeli Presence In Iraq." Mr. Tahsin alleged that there is a Jewish conspiracy to make another Jewish state in Iraq: "Is this the fulfillment of the Zionist dream to create a Jewish state from the Nile to the Euphrates?...Those who follow what is happening in Iraq will see that Israel has a hand in the country. Mossad, its secret service, has many operation centers both in Baghdad and in other major cities in Iraq; their job is to organize terrorist activities to guarantee that Iraq remains unstable..."

According to Iraq's Al-Zaman newspaper, a report on September 21 detailed that the Saudi Royal Family's Muslim World League "accused Zionist and Christian Evangelist organizations of infiltrating Iraq to introduce programs to Iraqi youth in order to weaken their Islamic and national sentiments."

However, the most popular conspiracy about Christians in Iraq has to do with the bombing of churches. In an October 18 article published by Iran's Mehr News Agency, Hassan Hanizadeh wrote, "The criminals who bombed the churches in Baghdad were trying to create a religious crisis in Iraq, and the Zionists definitely have great experience in creating religious conflicts...Bearing in mind the goals of the bombers, it can be concluded that these terrorists most likely have some connection with Israeli intelligence agencies."

A November 12 sermon by Hezbollah leader Sheikh Hassan Nasrallah on Al-Manar TV explained the CIA and the Mossad are responsible for church bombings: "Examine very carefully those who blow up the churches...You will discover that some of these groups are run by the CIA...the Mossad."

Writing in Al-Hayat on December 12, columnist George Haddad similarly explained, "During a few weeks only, a third wave of outrageous bombing of Christian churches occurred in Iraq...If we look at these crimes from a criminal investigation point of view, and ask who is benefiting from these attacks, it would not be difficult to discover that the American occupation, international Zionism, and Israel get the real benefits from such attacks."

In addition to blaming the CIA for attacking Iraqi churches, a posting that appeared on the Islamist Web site Islam Online on November 10 claimed that "the U.S. occupation troops are gassing resistance fighters and confronting them with internationally banned chemical weapons." The article added, "The U.S. troops have sprayed chemical and nerve gases on resistance fighters, turning them hysteric[al] in a heartbreaking scene...Some Fallujah residents have been further burnt beyond treatment by poisonous gases."

It should be noted that while conspiracy theories continue to be rampant in the Middle East, Arab reformists and progressives have spoken out against them. For more information on this rising phenomenon, visit MEMRI's Reform Project at www.MEMRI.org/reform.

Arab, Iran Press Blames U.S. for World Calamities

December 29, 2004

The year 2003 saw conspiracy theories from the Middle East surrounding most major events, ranging from the destruction of the Space Shuttle Columbia to the capture of Saddam Hussein. This year saw a continuation of this trend. The following is the fifth and final article in a series about conspiracy theories from the Middle East in 2004.

Previous articles in the summer and fall documented the practice of blaming America and Zionists for worldwide calamities, Egyptian conspiracies about September 11, and accusations about activities of Americans and Jews in Iraq. The following is a list of the top 10 conspiracies published in the Arab and Iranian press during the last months of 2004.

10) American and Zionist Control of Al Jazeera: Muslim critics of Al Jazeera frequently charge that it is run by America and Israel. In the Tehran Times on December 2, Hassan Hanizadeh wrote that "rumors arose suggesting that the network was established by U.S. and Israeli agents in order to present a bad image of Islam to the world...The actions of the network gradually revealed the fact that Al Jazeera officials, on the orders of Zionist agents, are trying to... tarnish the image of Islam."

9) Zionists Spreading Diseases Among Arabs: Hezbollah's Al-Manar TV broadcast a program in November about "Zionist attempts to transmit dangerous diseases like AIDS through exports to Arab countries." Iran's Mehr News Agency ran a story titled "Israel Exporting Carcinogenic Foodstuff to Arab States" on November 24, writing, "The trend has led to an increase in leukemia and liver and stomach cancer in the Arab countries which have imported Israeli foodstuff."

8) Iran Accuses Abu Musab al-Zarqawi of Working for Jordan: The Iranian Mehr News Agency alleged on December 26 that "Jordan sent Abu Musab al-Zarqawi to Iraq to carry out destabilization activities," and that Prime Minister Allawi is controlled by Jordanian officials. The report also accused Jordan of planning to resettle 5 million Palestinian Arabs in Iraq.

7) Jews Were Behind the September 1-3 Chechen Beslan School Attack: Dozens of religious, political, and press personalities in the Middle East have blamed the Beslan attack on Jews and Israelis. Ali Abdullah, a conservative Bahraini religious scholar, claimed on September 5 that Israelis were behind the attack, as part of an effort to "tarnish the image of Muslims."

6) Israel Stealing Body Parts of Palestinian Arab Children: Iran's Sahar 1 TV is currently airing a weekly series that premiered on December 13 called "Zahra's Blue Eyes." The show includes graphic scenes of Palestinian Arab children whose eyes have been surgically removed and stolen by Israel.

5) Jews Tamper with the Koran: The Kuwaiti English language daily Arab Times published an article on September 21 titled "New Education Plan 'Enforced;' Jews tamper with Holy Quran." The article reported on the dean of the College of Sharia and Islamic Studies of Kuwait University, Mohammad Abdul Razak Al-Tabtabaei, who had "warned citizens and residents of some versions which have been manipulated by Jews." Mr. al-Tabtabaei explained, "For long Jews have been plotting against Muslims using various means...They have been damaging the Holy book by the changing and deleting verses with the aim of hiding the truth."

4) America To Invade Pakistan and Annihilate Muslims: Maqdoom Mohiuddin favorably reviewed a book in the Saudi Gazette on October 19, which advances the theory that "there have been rumors about a possible invasion of Pakistan... The U.S. has started implementing its program...The book also reveals that the other possible targets of the U.S. will be Sudan, Indonesia, Nigeria, and Turkey." Mr. Mohiuddin claims that President Bush's true goal is the "total annihilation of the Muslims."

3) America/Jews Behind the October 7 Taba Attacks in Egypt: Adli Barsum of the Egyptian government daily Al-Gumhuriya is one of dozens of Arab columnists who propagated conspiracy theories about the Taba attacks. He asked in Al-Gumhuriya on October 12: "Who planned the bombings in Taba?...Is it the Mossad?...Is it the CIA?" The London Arabic-language daily Al-Quds Al-Arabi

reported that a headline in Egypt's weekly Al-Mustaqbal Al-Jadid, the official organ of the National Party's Policies Committee under the leadership of Gamal Mubarak, read "Proof of Mossad Involvement in the Explosions in Taba."

2) Israel Killed Yasser Arafat: Writing in the Egyptian weekly Al-Ahram, Ghada Karmi claimed on November 11, "A close confidant of Ariel Sharon has stated that the latter had 'eliminated' Yasser Arafat 'through his cooks.'" On November 11, Khaled al-Batesh, leader of Palestinian Islamic Jihad, told United Press International that "during Sharon's last visit to Washington, he told U.S. President George W. Bush, 'We will help God to get rid of him.'"

1) American Soldiers Stealing Organs from Iraqis: In the Saudi daily Al-Watan on December 22, an article by Fakhriya Ahmad alleges that, based on secret European military reports, American personnel in Iraq are stealing human organs: "The reports confirm the finding of tens of thousands of mutilated cadavers missing parts...These teams offer $40 for every usable kidney and $25 for an eye." The report also appeared in the Syrian daily Teshreen and in Iran's Jomhour-ye Islami. As 2004 comes to an end, Arab and Iranian conspiracy theories were again rampant. This follows an ongoing trend, which will be covered in this column in 2005.

2005

Saudi Arabia's Terror Conference, Part I

January 5, 2005

Since 9/11, the Saudi royal family has spent millions of dollars to improve its image. This includes a recent ad campaign in The New Republic, which are "Sponsored by the People of Saudi Arabia – Allies Against Terrorism." Another full-page ad appeared in the USA Today, stating, "For far too long, rumors have been accepted as truth... The 9/11 Commission Report finally reveal[ed] the facts." However, despite the Saudi PR efforts which claim that the report completely vindicates Saudi Arabia, it in fact states that Saudi Arabia is "a problematic ally in combating Islamic extremism" and that "significant problems remained" regarding its role in the war against terror.

Another attempt by the Saudis to improve their image is the convening of an international conference against terrorism to be held in Riyadh on February 5-8. According to Saudi government sources, 43 countries will attend. Prince Turki Ibn Muhammad Bin Saud Al-Kabir, an assistant undersecretary at the Foreign Ministry said, "We have invited all countries that have suffered from terrorism ... and all have agreed to take part."

The conference was initially announced at the UN General Assembly in September by Saudi Assistant Minister of Foreign Affairs Mizar Midani. When asked why Israel was not invited, he accused the Jewish state of "being responsible for extremism in the region."

Meanwhile, the Iranian Republic News Agency reported that Iranian President Muhammad Khatami will be attending. Prince Al-Kabir, who is the chairman of the conference, told the Saudi Gazette on November 3 that a large contingent from the U.S. will also attend.

Whereas the conference's stated goal is to "eradicate the roots and causes of terrorism," the Saudi royal family's government has in fact done very little to pursue this – outside of the Kingdom. The Saudi war on terror has exclusively focused on fighting only the wing of Al-Qa'ida within Saudi Arabia. As Saudi writers, TV commentators, professors, clerics, and members of the royal family often explain, Jihad is acceptable as long as it is not within or against the Kingdom.

Writing in the Saudi daily Al-Riyadh on April 26, 2004, Abdul Waheed Al-Humaid referred to the attacks of that week in the Saudi capitol Riyadh as unjustifiable terrorism. He explained, however, "if there are people who want to wage Jihad and fight the enemy, there are more than a thousand [legitimate] ways to do so."

Saudi Sheikh Abdallah Al-Muslih, Chairman of the Commission on Scientific Signs in the Koran, Sunna of the Muslim World League, and former Dean of Islamic Law in the Saudi city of Abha appeared on Iqra TV on May 20th, stating that Jihad - inside the Kingdom - is not allowed. He also addressed the current debate amongst leading Saudi clerics about suicide bombings against U.S. troops. He cited teachings from Islamic history giving precedents to such actions that as long as soldiers from Dar Al-Harb (countries outside Muslim rule) are targeted, "there is nothing wrong with suicide attacks if they cause great damage to the enemy." Al-Muslih ended by emphasizing, "[When] we speak of [attacks] in Muslim countries, such as Saudi Arabia ... this is forbidden... This is the land of the Muslims. We must never do this in a Muslim country."

Saudi support of Jihad outside the Kingdom and against U.S. troops was recently the subject of a fatwa by 26 leading Saudi religious scholars from the most prominent universities in the Kingdom. According to the fatwa, released in November, killing U.S. soldiers in Iraq is allowed. The fatwa, which came one month before the suicide attack by a Saudi bomber on an American mess hall in Mosul which killed 14 U.S. soldiers, stated: "Fighting the occupiers is a religious duty... It is a jihad to push back the assailants... Resistance is a legitimate right."

The Saudi embassy in DC tried to distance itself from the fatwa. However, according to Saudi law, the government is the only body that can lawfully issue such a fatwa. The unauthorized religious authorities who sanctioned the killing of U.S. troops have yet to be punished.

Saudi Prince Amr Al-Faisal responded to the fatwa in Arab News on December 6 by explaining that the 26 scholars don't represent official Saudi policy, but that they do represent "a significant opinion in the Muslim world on the proper manner for dealing with foreign occupiers in Iraq." Prince Amr added, "U.S. forces in Iraq are not a group of friendly boy scouts out to help elderly Iraqi ladies cross Baghdad streets."

In the coming weeks, this column will expose Saudi Arabia's continual embrace of Jihad, as well as its support of a culture of hatred against the West, Christians, and of course Jews. Unless the Saudi terrorism conference addresses these issues, the upcoming columns will explain why the U.S. should not take part in the Saudi Arabia's February conference on terrorism.

Saudi Arabia's Terror Conference, Part II

January 12, 2005

"This perverse ideology [Wahhabi Islam] has spread all over Saudi Arabia, in the schools, the mosques...satellite channels...There's a videotape now circulating in Saudi Arabia of a boy age 10 or less [in a Saudi orphanage]. He is asked, 'who is your role model?' And he answers, 'Osama bin Laden.'" This damaging statement was not spoken by an opponent of the Saudi regime, but by Saudi Prince Khaled al-Faysal on Al-Arabiyya TV on July 14.

As part of MEMRI's TV Monitoring Project (www.memritv.org), Saudi government-controlled TV channels are continually monitored. These channels include shows with leading Saudi religious figures, professors, members of the royal family, government leaders, and intellectuals. Constant themes include calls for the annihilation of Christians and Jews, rampant anti-Americanism and anti-Semitism, and support of jihad.

With the Saudi government preparing to convene an international counter-terrorism conference February 5-8, which will include participants from leading state-sponsored terrorism regimes such as Iran - as well as reports by the Saudi Foreign Ministry that America is sending a large delegation - it is important to review the content of the Saudi press, particularly TV, as it relates to terrorism and hatred of non-Muslims.

A recent example of hatred of Christians and Westerners is Saudi TV's coverage of the tsunami disaster. The Saudi Embassy in D.C. boasted about the numerous telethons that were held to support victims of the natural disaster. However, these telethons included influential Saudi figures who explained that the celebration of Christmas and New Years is what led to the death of over 150,000 "infidels."

Ibrahim al-Bashar, an advisor to the Saudi Justice Minister, spoke on Al-Majd TV on January 5 and said that the countries hit by the tsunami were being punished by Allah for lying, sinning, and being infidels. The Saudi sheikh Fawzan Al-Fawzan, who taught at the Saudi Supreme Court of Justice, also said to Al-Majd TV on December 31 that Allah's destruction of these countries was timed to coincide with Christmas - a time of rampant fornication and perversion. Saudi cleric Muhammad al-Munajjid, Imam of the Umar ibn Abd al-Aziz Mosque in Khobar, claimed first to Al-Majd on January 1 that Allah took revenge on the "criminal's" celebration - the Christmas holidays - and elaborated on January 6 that "perversions" during the Christmas holidays were to blame.

Sermons from Saudi mosques frequently contain calls to fight non-Muslims. In a sermon from Medina broadcast on Saudi TV channel 2, Sheikh Saleh Bdeir said on June 25: "The enemies of Islam, the Jews, Christians, atheists, and those from among the infidel Westernized who are enslaved by them within the Muslim community, never cease attacking the Islamic nation." He called to Muslims, "Confront your enemies' efforts with greater and stronger efforts, before these enemies become stronger."

In the early summer, Saudi Sheikh Ahmad bin Abd Al-Latif, a professor at Um Al-Qura University, was asked on Saudi TV if it is permitted under Islamic law to pray for the annihilation of Christians and Jews. He answered, "Cursing the oppressing Jews and the oppressing and plundering Christians and the prayer that Allah will annihilate them is permitted."

Saudi cleric Aed Al-Qarni spoke on Saudi Iqra TV channel on December 12 about why Jews and Christians will burn in Hell: "The Jews take pride in something they lie about; the Jews and the Christians...They say: 'Oh people, we, the Jews and Christians, are the sons of Allah...' They are lying, [may] Allah's wrath [be] upon them...If you are truthful, will Allah burn you in hell for your sins?...You will be punished for your lies."

Calling for the throats of Christians and Jews to be slit and their skulls shattered, Mr. Al-Qarni told Iqra TV on December 26: "We Muslims should be rebuked. One billion 200 million...are incapable of taking action...of harming the Jews...I pray to Allah that He will make the enemies fall...and that He will destroy the Jews and their helpers from among the Christians...We curse them...and pray that Allah will annihilate them, tear them apart, and grant us victory over them...Throats must be slit and skulls must be shattered. This is the path to victory, to shahada..."Citing a hadith in a lecture on January 9 that aired on Iqra TV, Mr. Al-Qarni explained that Jews, "the brothers of apes and pigs," and Christians should not be slaughtered only if they convert to Islam.

Clearly, in Saudi Arabia there exists a culture of hatred against the West and non-Muslims. This is the same culture that produced Osama bin Laden and 15 of the 19 September 11, 2001 hijackers. The Saudi royal family insists it is fighting hatred - but anyone watching the TV stations under their absolute control will find they are not telling the truth. The examples mentioned in this article are only a few of the many that appear daily to incite their viewers to violence.

In the coming weeks, this column will detail the Saudi perception of "the Jews," as well as look at how the country supports jihad in Iraq.

Saudi Arabia's Terror Conference, Part III

January 21, 2005

As the February 5-8 Saudi conference on terrorism approaches, it is worth examining the Saudi royal family's position on jihad against Jews, as well as how this relates to Israel's exclusion from the conference. The Saudi assistant minister of foreign affairs, Mizar Midani, explained that Israel is excluded because it is "responsible for extremism in the region."

One of the most common forms of anti-Semitism embraced in the Middle East, including Saudi Arabia, is the phenomenon of blaming everything on Israel, Zionists, and Jews (and in recent years America, too). As Sheikh Abd Al-Qader Shiba Al-Hammad stated on Saudi MBC TV on August 6: "There is no evil in the world that the Jews are not behind...In the depths of history and in the future...you will find that the finger of the Jews...is behind everything."

The magazine Al-Jundi Al-Muslim, or the Muslim Soldier, published by the Religious Affairs Department of the Saudi armed forces, released an article on May 1 in its Know Your Enemy section that stated: "The majority of revolutions, coups d'etat, and wars which have occurred in the world...are almost entirely the handiwork of the Jews...in order to implement the injunctions of the fabricated Torah, the Talmud, and the Protocols [of the Elders of Zion], all of which command the destruction of all non-Jews in order to achieve their goal - namely, world domination."

Blaming the Jews/Zionists/Israel for the world's ills begins in the upper echelons of Saudi society. The interior minister, Prince Nayef, publicly blamed September 11 on "the Zionists." Following terror attacks in Saudi Arabia last year, Crown Prince Abdallah blamed "Zionists," a position that was supported by members of the royal family. Prince Nayef explained at the time: "Israel and Zionism are behind Al-Qaeda."

Given the fact that Saudi Arabia has never recognized Israel, it is not surprising that the Saudis have a long history of hatred of Jews and support for terrorist attacks and jihad against Israel. It has been documented by official Saudi government sources that committees created by leading princes, such as the Popular Committee for Assisting the Palestinian Mujahideen and the Support to the Committee for the

Al-Quds Intifada, which are now the subjects of numerous lawsuits from victims of terrorism, have given more than $4 billion to "mujahideen fighters" and "families of martyrs."

The daughter of King Saud, Princess Fahda, is the historian of her father's reign and wrote an article last year that stated "...King Saud made the right diagnosis: 'The Zionist threat is like cancer - in dealing with it neither medicine nor surgery will do any good.' This royal statement was meant to emphasize that the Arabs do not, and will not, accept an Israeli state amidst them.'" She added that under the leadership of King Saud, the Saudi representative to the United Nations called on it to establish an agency "to help resettle Jews [now in Israel] in their former European homes."

The Saudi Abdallah Al-Quba wrote in the journal Saudi Foreign Policy in 1986 about the official Saudi policy of jihad against Israel: "We will not be blamed if we undertake the method of holy jihad. The racist military religious war that Israel wages...will not come to an end but by jihad. The meaning of holy jihad is a sole... Islamic confrontation, where we would replace our entire spiritual, educational, diplomatic...and military capabilities ...into it, for a long persistent jihad."

All aspects of Saudi culture, including schoolbooks, sermons, press, and government institutions espouse hatred of the Jews. In a show on Iqra TV on August 26, a reporter on the streets asked passersby: "Would you shake hands with a Jew?" Answers included: "No, because the Jews are eternal enemies," and "Of course not, so I wouldn't have to consider amputating my hand afterwards." Next, the interviewer asked "If a child asks you who 'who are the Jews,' what would you answer?" The interviewees responded: "The enemies of Allah," and "Allah's wrath is upon them, as the Koran says...They are the filthiest people on the face of this earth...The solution is clear...If only [the Muslims] declared jihad."

A statement posted on the Saudi embassy in Washington, D.C.'s Web site detailed that the Saudi terror conference will focus on "the concepts and causes of terrorism." It was also reported by Saudi officials that all countries affected by terrorism will be invited. Israel's exclusion could be because, according to Saudi law, no Jews are permitted to set foot inside Saudi Arabia, as the Saudi Ministry of Tourism Web site stated last year. It also could be due to the fact the Saudi royal family has always supported jihad against Israel, and as many of its princes have repeatedly stated, suicide attacks against Israel are legitimate acts of resistance, not terrorism.

One would hope that the State Department is aware of the Saudi position calling for jihad against Jews and that any American participation at the conference may unwittingly imply support of this policy.

Arab Holocaust Denial 60 Years Later

January 24, 2005

To mark the 60th anniversary of the liberation of the Nazi concentration camps, the U.N. will hold a special session today. Over 100 nations including Arab states such as Egypt and Saudi Arabia reportedly support holding the session.

At last count, Syria was not among them. To understand the official Syrian position on the Holocaust one needs to read what is said in its government-controlled press. Discussing the "myth" of the Holocaust, in the September 6, 2000, edition of the Syria Times, Mohammad Daoud wrote: The "most famous myth is that of the so-called Holocaust ... we strongly believe that gas chambers were not used for burning of Jews."

Arab religious leaders also frequently espouse Holocaust denial. The Saudi sheikh Adel Bin Ahmad Banama said at a mosque in Jeddah on October 22, 2000: "Today, they [the Jews] disseminate everywhere the lie of the Holocaust and claim that Hitler killed 6 million Jews in gas chambers ... this is pure falsehood." The mufti of Jerusalem, Ikrima Sabri, is imam of Al Aqsa mosque and the highest religious authority in Palestinian Arab Islam. Mr. Sabri gave an interview to the Italian newspaper La Republica on March 24, 2000, and stated: "Six million Jews dead? No way... Let's stop with this fairytale."

Palestinian leader Mahmoud Abbas's 1982 doctoral dissertation was based on Holocaust denial. In the introduction to its first Arabic edition, well-known Holocaust deniers are referenced, the total number of Jews murdered is formulated at "less than a million," and doubts are raised that gas chambers really existed.

To prove such claims, Arab press outlets often cite theories of European Holocaust deniers, who are celebrated in the Middle East. Fatma Abdallah Mahmoud wrote in Egypt's Al-Akhbar on April 29, 2002: "With regard to the fraud of the Holocaust ... many French studies have proven that this is no more than a fabrication, a lie, and a fraud!" In an article in Egypt's Al-Gumhuriya on March 4, 2000, columnist Lutfi Nasef wrote: "Zionist propaganda continues ... to raise the issue of the Nazi crematoriums for Jews, although [European] historical evidence ... proved that claims that such crematoriums existed in the Nazi detention camps are jokes."

Columnist Muhammad Abd Al-Azim also in Al-Gumhuriya on February 22, 2000 explained that "Israeli claims 6 million Jews were killed" were discredited by infamous French Holocaust denier [and convert to Islam] "the great French intellectual, Roger Garaudy." Another European Holocaust denier, David Irving, was praised by Isam Al-Khadhr, a columnist for the Syrian paper Al-Baath, on September 7, 2000: "The English Historian David Irving's description of the so-called Jewish Holocaust by the Nazis in gas chambers as a big lie was enough to shake the foundations of Europe and turn Irving into a target for a campaign of slander ... the caliber of the 'Holocaust' lie, the greatest lie in History."

In the last week of December 2004, interviews with two infamous Holocaust deniers appeared in the Iranian press. French professor Robert Faurisson, a former lecturer at Lyon University, gave an interview to Iran's Mehr News Agency on December 18 discussing "The big lie of the alleged Holocaust." On December 29, Australian Fredrick Toben also gave an interview to Mehr stating: "No Arabic speaking regime, except recently Egypt, has openly pushed for adopting a 'Holocaust' expose. It is this information that will help to dismantle the Zionist entity because the State of Israel is founded on the 'Holocaust' lie."

Mr. Toben is referring to the activity of some of Egypt's leading think tanks. For example, on June 24, 2004, Rifat Sayyed Ahmad, director of the Jaffa Research Center in Cairo published an article titled "The Lie About The Burning of the Jews" in Egypt's ruling National Democratic Party's paper. The article stated: "Did this Holocaust indeed take place, and what is the truth about the numbers? ... What interests us here is that this lie [about] the burning of the Jews in the Nazi crematoria."

To date, only a handful of prominent Arabs have spoken against Holocaust denial. President Mubarak's political adviser, Osama Al-Baz, has done so on numerous occasions. But this usually results from criticism of Egypt for anti-Semitism in its press. With the majority of Arab states supporting today's U.N. special session, one can hope that this marks some aspect of change in the way the Arab world relates to supporting anti-Semitic themes such as holocaust denial.

Saudi Arabia's Terror Conference, Part IV

January 26, 2005

With the February 5-8 Saudi counter-terrorism conference approaching and America reportedly planning to participate, it is important to understand the Saudi position vis-avis jihad against American troops in Iraq. While Saudi officials proclaim that Saudi Arabia does not support this jihad, a look at Saudi government press coverage and statements by religious leaders and Iraqi government officials proves otherwise.

A professor of Islamic law at Saudi Arabia's Um Al-Qura University, Yassin Al-Khatib, declared on Saudi/UAE Al-Majd TV on May 10 that, "the fact that [America] entered [Iraq] ... makes it every Muslim's duty to go out against them, not only the Iraqis ... jihad today has become an individual duty that applies to each and every Muslim ... When the Muslims fought in Afghanistan they destroyed the Soviet Union ... It collapsed, and Allah willing so will [America] collapse."

Many Saudis have been listening to religious figures such as Mr. Al-Khatib, who use the Saudi press to call upon Muslims to fight America in Iraq. This has been confirmed on numerous occasions by Iraqi press and government officials. The Baghdad newspaper Al-Mashriq reported on January 12 that the Iraqi Ministry of Human Rights published the names of 14 Saudis caught planning acts of terrorism. On November 1, Prime Minister Allawi announced the arrest of a group of Saudi insurgents. On October 3, the Iraqi interior minister, Fallah Al-Naqib, told a press conference that many insurgents killed in the previous weekend's fighting in Samarra were Saudis. On July 29, the London daily Al-Sharq Al-Awsat announced that Iraqi police arrested the head of one of Baghdad's terrorist groups - a Saudi.

According to a December 15 article in the Saudi Gazette, growing numbers of Saudis went to fight American troops following an internal crackdown against terrorists targeting the Saudi government. Having been spurred on by Saudi religious leaders, many reportedly went to fight American troops in Fallujah. For example, Saudi cleric Aed Al-Qarni stated on December 26 on Saudi Iqra TV: "The jihad in Fallujah is a source of pride. It is astounding: a city of 250,000 residents opposing a superpower ... downing their planes, destroying equipment, slaughtering them, taking them hostage, and proclaiming 'Allah Akhbar' ... Who can say even one word against this true jihad against these colonialist occupiers?"

Dozens of reports in the Arab press have detailed the stories of individual Saudis going to martyr themselves by waging jihad in Iraq. For example, "Abdullah," a Saudi fighter, was interviewed by the London daily Al-Hayat last year. He said he came to Iraq out of a religious obligation to wage jihad against "American aggression ... When I left Saudi Arabia, I was ready to become a martyr." A recent well-known instance of a Saudi who martyred himself is the suicide bomber who killed 22 people in the dining hall of an American Army base in Mosul, Ahmad Al-Gamadi, who had been recruited in Saudi Arabia. Al-Sharq Al-Awsat reported that he is just one of hundreds of Saudis who have been recruited to fight American soldiers.

The families of Saudi jihadis have expressed pride over the martyrdom of their children. Al-Hayat reported last year about two families in eastern Saudi Arabia. The father of Suheil Al-Sahili, who was killed in northern Iraq, told the paper: "I thank Allah that [our son] attained what he sought. For 14 years he sought [martyrdom]. He always pointed to his head and wished that a rifle bullet would split his forehead ... We got a phone call from him finally, in which he said he was going to the jihad in Iraq." The brother of Abd Al-Hadi Al-Shehri told the paper: "From a young age he wanted jihad ... After fulfilling this commandment of pilgrimage to Mecca, there was no contact with him until news of his martyrdom reached us." Another article from the same paper on April 12, 2004, reported that a Saudi national was "congratulated" by many for the death of his son as a martyr in Fallujah.

Over the past month, this column has been trying to expose Saudi Arabia's hypocrisy in claiming that it is vigorously opposing jihad and anti-Western incitement, when in reality hatred of America and support for jihad against it is a mainstream Saudi view propagated by the government's press and religious establishment. The upcoming Saudi terror conference must be viewed with suspicion, and as part of the ongoing game the kingdom has been playing regarding terrorism since before September 11, 2001.The Saudis claim that they are opposing hatred and jihad, when nothing could be further from the truth. Any country participating in the terror conference, including America, must understand that is only helping the Saudis continue to play their game.

The Iranian Threat: Ayatollah Ali Khamenei

February 9, 2005

Ayatollah Ali Khamenei is the "supreme leader" of Iran and commands the Iranian military. He is also at the forefront of supporting jihad against America. Many of his sermons are broadcast live on Iranian state-controlled TV.

The supreme leader frequently warns America against interfering with Iran's nuclear program. On July 5, 2004, he said to a crowd chanting "Death to America," " "We will cut off the hand that is sent to invade and work against our people's interests ... If the enemy has the audacity to harm and invade, our blows against it will not be limited to the borders of our country."

This past summer, Mr. Khamenei gave a speech broadcast on the Iranian channel Jaam-E-Jam 2 about the same subject: "The enemies of Iran - led by the most despicable of all, the U.S. government ... opened their filthy mouths ... We have no need for a nuclear bomb ... The Iranian people have been defeating America for the past 25 years. The world of Islam has been mobilized against America for the past 25 years. The people call 'death to America!'"

In response to President Bush's State of the Union address on February 3, Mr. Khamenei called America "one of the lands of a Seven-Headed Dragon." As the main propagator of anti-Americanism, the supreme leader often compares Mr. Bush to Adolf Hitler and America to Nazi Germany. An article was devoted to this subject in the Persian daily Jomhour-e Eslami, paraphrasing comments made the week of August 30, 2002, by Mr. Khamenei: "The language of the U.S. president resembles the language of Hitler. ... If Hitler had to show the face of a bloodthirsty dictator, he would have had to adopt [the face of] George W. Bush."

The article continued, describing Mr. Khamenei's philosophy that America is the modern incarnation of Nazi Germany: "There is a great resemblance between the behavior of today's Americans and the behavior of the Nazis. ... The Americans are infected today with satanic pride and arrogant egotism ... Bush is heading towards a dead end on the same path on which the dictators who caused the two world wars."

Mr. Khamenei is confident that America will soon collapse. He was quoted in Jomhour-e Eslami on May 20, 2004, stating that "the end of the U.S. will begin in Iraq. As the imam said, 'One day the U.S., too, will be history.' In light of what happened in Iraq, we can see now that he is right, because such events move the U.S. down the slope, and they will taste the bitterness of sure defeat."

One of the supreme leader's weapons to be used to defeat America is jihad by students who are willing to become martyrs. Speaking to a group of students in the first week of 2005, the Iranian newspaper Sharq quoted Mr. Khamenei calling on them to embrace jihad: "The enemies of Iran tried to humiliate and diminish the value of martyrdom and the culture of jihad in the eyes of the youth, particularly students. ... This is the same wondrous element that gives those fighting jihad for the path of truth the strength to overcome any scheme of the front of hostility."

On January 20, urging students to continue to promote the culture of jihad and martyrdom amongst themselves, he stated: "When we encounter the name of a student who committed martyrdom we are confident that the acceptance of martyrdom and of the jihad that led to this martyrdom stemmed from self-awareness. ... This intensifies the value of the act. Sanctify and praise your exalted martyrs."

In an address broadcast on Iranian TV on October 22, 2004, Mr. Khamenei explained: "Jihad means to strive for the lofty values. ... Whoever fights for these values ... Allah's angels point to him in heaven. Do not consider those who are slain in the cause of Allah as dead. ... When this great nation, with its zealous and faithful youth ... resist the enemy ... it will let it taste defeat once more."

Threats against America by leading Iranian officials have become a common occurrence. Over the next few months, this column will focus further on them.

Saudi Arabia Tops List For Anti-Semitism

February 23, 2005

In the State Department's inaugural annual Report on Global Anti-Semitism released last month, a total of 25 examples of anti-Semitism were cited from the Middle East. One third were based on research from Memri's Anti-Semitism Documentation Project.

Egypt was criticized for an increase in anti-Semitism in its media. One example cited is a March 18, 2004 article by Abd Al-Wahhab'Adas, deputy editor of Al-Jumhuriya, who accused Jews of perpetrating the Madrid terrorist attacks. Also noted was the placing of "The Protocols of the Elders of Zion" in the Alexandria Library in December 2003 in an exhibit on Jewish holy books. The section on Lebanon criticized "Al-Shatat," a television series based on "The Protocols of the Elders of Zion" broadcast in 2003 by the Lebanese terrorist organization Hezbollah's Al-Manar TV. Syrian government ministries were also cited for involvement in creating the series, and the Syrian regime was condemned for ongoing anti-Semitism in its media.

The United Arab Emirates was praised in the report for closing the Arab League think tank called The Zayed Center, in August 2003. It was noted that it was closed due to the publication of reports with anti-Jewish themes, and for hosting speakers who promoted these views. In response to the State Department, the UAE's deputy prime minister, according to the Khaleej Times, "accused the authors of the report of choosing to mimic ... charges made by MEMRI."

Some critics of the report noted that the section devoted to the Middle East was mostly a whitewash of the rampant hatred of the Jews in the Arab world. The following are some notable examples, which could have been included:

* In Egypt, Hussam Wahba, a columnist for the religious Egyptian weekly magazine Aqidati, published by a foundation linked to Hosni Mubarak's National Democratic Party, wrote an article on August 10, 2004, based upon blood libels and accusing Judaism of promoting ritual murder: "The Talmud also says 'Kill anyone who is not Jewish, even if he is pious ...' The Talmud says that 'it is righteous for a Jew to kill a non-Jew with his own hands, because whoever kills a non-Jew is offering a sacrifice to God.'"

* In Lebanon, leading clerics and government officials commonly express anti-Semitic statements. Al-Manar TV hosted the secretary-general of the Islamic Universities' Association, Dr. Ja'far Abd Al-Salim, on April 22, 2004, for a discussion about Jews. Sheikh Sabonji explained: "Those responsible for all civil strife and other problems throughout history were the Jews. This is verified by anyone who has read Jewish literature and history." He added: " 'The Protocols of the Learned Elders of Zion' clearly refer to this. All the pornographic films in the world are made by Jewish companies, and the same goes for drugs."

* Saudi Arabia should be considered the most anti-Semitic nation in the region. The report stated, "there were frequent instances in which mosque preachers ... used strongly anti-Jewish language in their sermons," but mentioned a "decline in frequency" in anti-Jewish language in sermons. There was, however, no decline in hatred of the Jews expressed in religious programs on Saudi government TV (to view, visit www.memritv.org).

Saudi government publications also continue to espouse hatred of Jews. For example, a journal titled Al-Jundi Al-Muslim ("The Muslim Soldier"), which is published by the Religious Affairs Department of the Saudi armed forces, released an article on May 1 in its "Know Your Enemy" section. The article was titled "The Jews in the Modern Era," and here's a passage: "The majority of revolutions, coups d'etat, and wars which have occurred in the world ... are almost entirely the handiwork of the Jew ... in order to implement the injunctions of the fabricated Torah, the Talmud, and the 'Protocols [of the Elders of Zion'], all of which command the destruction of all non-Jews in order to achieve their goal - namely, world domination."

While there was criticism of this report by organizations fighting anti-Semitism, it was an important first step in monitoring incitement against Jews in the Middle East. Unfortunately, statements by leading Arab figures, such as mufti Ikrimeh Sabri on Al-Majd TV last week: "Anyone who studies 'The Protocols of the Elders of Zion' and specifically the Talmud will discover that one of the goals of these protocols is to cause confusion in the world" point to the fact that next year's report will also have countless examples of anti-Semitism to choose from within the Middle East media.

Iran Leaders' Threatening Rhetoric

March 3, 2005

"If you [Americans] behave with disrespect even just a little bit, [the Iranian people] will punch you in the mouth so hard that all your devouring teeth will fall out." So warned Iran's Ayatollah Mohammad Ememi-Kashani in a Friday sermon at Tehran University, which aired on January 28, 2005 on Iranian TV's Channel 1.

This statement, and countless other threats by leading Iranian political, religious, and military leaders, which are frequently made on Iranian government-controlled TV channels, will be shown today at the U.S. Capitol as part of Memri's lecture series.

Programming on Iranian government-controlled TV includes a complex propaganda campaign against America. For example, the Iranian News Channel Irinn TV covered a conference titled "The World Without America" held by the Islamic Union of Students in Iran on November 8, which sought to "clarify what is going to happen in the near future." The head of the Iranian Islamic Student Union, Hamid Aqaii, told the channel, "We believe that according to the predication of the late imam [Khomeini] and the honorable leader [Khamenei], signs of the decline and fall of America have begun to appear."

Anti-American music videos are an important part of Iranian propaganda. One video appearing on Sahar TV on August 12, was titled "Satan." Images of President Bush, Secretary of Defense Rumsfeld, Colin Powell, the former secretary of state, and Congress were shown with the words, "This is Satan, the source of tyranny."

Another anti-American music video which appeared on Irinn TV on August 12, produced by the Iranian Revolutionary Guards, was titled "The Scheme of Thais," referring to a Greek courtesan who joined Alexander the Great and persuaded him to set fire to Persepolis. The video shows the Statue of Liberty as the incarnation of Thais, calling her a "she demon" and "the anti-Christ." The Statue of Liberty morphs into a skeleton with American flag eyes, with a voice in the background singing: "America, Enemy of God ... America the mad demon." As the American flag burns, the song continues, with the words "light the pure flames to destroy the serpent ... America is the enemy of God ... America is the mad demon ... O men of the seven continents, awake! Awake! Light the pure flames to destroy the serpents; This impure octopus with seven heads ... must be killed."

Iranian TV devotes extensive coverage to personal attacks on Mr. Bush, often comparing him to Hitler. This summer and fall, a series titled "The New Fascism" was aired on Irinn TV. Images from the series depict Mr. Bush with a swastika in the background. Iranian strategic expert Ali Askari was interviewed about the series on September 14, and stated, "We have seen that 9/11 was a domestic need of the American administration. ... Either the Americans were involved in this matter, or they let the events of 9/11 develop as they developed." Iranian political expert Manouchehr Mohammadi added, "There are many events like these in American history. It has been proven that the Japanese military attack on Pearl Harbor ... was planned ... with the authorization ... of U.S. President Roosevelt. ... They needed a pretext such as this to enter WWII. ... The exact same thing happened on 9/11.This is not used only by Hitler and George Bush."

To mark the 26th anniversary of Iran's Islamic Revolution, on February 10, a contest covered by Irinn TV was held in Tehran, to see who could burn the best effigy of Mr. Bush. It was part of an anniversary parade, and as the reporter covering it explained, "this contest was held in response to the U.S. president 's babbling." Effigy makers were shown presenting their effigies to judges, while those in the audience yelled "Death to America."

When viewing Iranian TV, one gets the impression that the Iranian governing mullahs are mobilizing for confrontation with America. For instance, on February 10, on Jaam-E-Jam1, the deputy head of the Iranian Revolutionary Guards, General Zol-Qadr said: "The events in the Middle East ... were a great opportunity for us to get to know the American methods, strategies, doctrines, and tactics....

Americans are our greatest security challenge. Thus, we have been able to prepare for a confrontation with the American threat. We even know the American commanders' state of mind, and this is very influential in wars. ...There is no threat that our armed forces cannot identify and confront with all their might."

Iranian threats against America including terror attacks on American soil have been detailed in past columns. These threats continue almost daily on Iranian TV.

False Hopes For Assad Are Dashed

March 9, 2005

Since inheriting the Syrian presidency in June 2000, Bashar Assad has given dozens of interviews to the international press. Using the opportunity to let the world know what is on his mind, the former ophthalmologist's interviews with the press usually focus on issues such as Syria leaving Lebanon, his country's support of terror organizations including Hezbollah, his views on Al-Qaeda, and allowing Palestinian Arab terror groups to operate from Damascus.

In an interview with the editor of the Kuwaiti daily Al-Anbaa on May 25, 2003, Mr. Assad explained in what circumstances Syrian troops would leave Lebanon: "Had the calls [against the Syrian occupation of Lebanon] constituted a majority, we would not remain there for even a single day ... Had the majority seen the Syrian [military] as [an] occupier - I would leave. No army can remain against the will of the citizens. The Lebanese people do not accept [the term] 'occupation' in its literal meaning."

On Hezbollah, he told the Arab satellite news channel Al-Arabiya on June 9, 2003, "We said that Hezbollah is Lebanese, not a Syrian party ... They have their beliefs and we have ours. When our beliefs intersect, we go with them, and they with us. When there are disagreements, we remain silent ... We have no control over Hezbollah, other than agreement about the right of resistance." He also explained that the Americans "are always confusing Hezbollah with the Palestinian organizations, and sometimes with Al-Qaeda. The information they have is not accurate."

The issue of Palestinian terror organizations operating from Damascus is part of the reason Syria has been labeled a state-sponsored terror regime by the State Department. Most recently, groups operating from there were accused of planning the Tel Aviv terrorist attack of February 28, 2005. As President Bush stated at the National Defense University yesterday, "America and other nations are also aware that the recent terrorist attack in Tel Aviv was conducted by a radical Palestinian group headquartered in Damascus. Syria ... has a long history of supporting terrorist groups ... The time has come for Syria ... to stop using murder as a tool of policy, and to end all support for terrorism."

Regarding American pressure to close these offices, he told Al-Arabiya, "We said that these organizations, which recently arrived, such as Hamas and the [Islamic] Jihad, are conducting informational activities. There are Palestinian organizations that have been in Syria for decades ... But these organizations are conducting informational activities: They take responsibility for acts or express opposition to particular initiatives. But their true existence is in Palestine, on the West Bank and Gaza. Those truly responsible are on Palestinian land ... We told the Americans that closing the offices would not solve the problem."

On January 19, 2004, the London Arabic daily Al-Sharq Al-Awsat published a interview with Mr. Assad in which he further discussed the activity of Hamas and Islamic Jihad operating from Damascus: "When the Americans asked us to expel Palestinian elements, we answered in the negative, and clarified that they had broken none of our laws and done nothing to justify [expulsion]."

The Syrian leader also gave a wide-ranging interview at his home, "The People's Palace" in Damascus, to Al Jazeera in April 2004. On his views regarding Palestinian Arab terror organizations, he explained: "The Palestinian resistance is legitimate because it is popular ... Are all the hundreds, thousands, and millions who are resisting occupation in various ways, not necessary by military resistance, all members of the so-called Al-Qaeda - and we do not know if there exists anything called Al-Qaeda! ... What is happening with regard to popularity gives legitimacy to the resistance."

Regarding Osama bin Laden's organization, Mr. Assad asked, "Is there really such an organization?" during his May 25, 2003 interview with Al-Anbaa, "I cannot believe that bin Laden is managing to prevail over the entire world. All the countries in the world, headed by the U.S., fight against bin Laden. He cannot talk on the telephone or use the Internet, but he manages to contact the four corners of the earth. This is illogical and unrealistic. How can he possibly plan this way, and how does he succeed in acting? With regard to Al-Qaeda, is there really such an organization? It was in Afghanistan; does it still exist?"

When Mr. Assad was handed the Syrian throne following the death of his father, the West had high hopes that he would improve the situation in Syria. Many have been disappointed.

Lebanon's Tipping Point

March 16, 2005

Following the massive pro-Syrian Hezbollah sponsored rally last week many in the Western media began to take a second look at the terror organization and its impish leader Sheikh Hassan Nasrallah. Robert Novak, appearing on CNN's Capital Gang on Saturday night quipped, "I'm glad we're [the U.S. is] dealing with them."

In Lebanon, the mainstream English daily, The Daily Star had an editorial on March 7, that stated, "Hezbollah is not a problem – it is part of Lebanon's solution." An editorial in Iran's Tehran Times declared, "extra-regional powers could no longer leave Hezbollah out of international political calculations... which is growing every day... and is able to influence all ... international political developments."

Not everyone has fallen in love with Hezbollah. Last week the European Parliament labeled it a "terrorist" organization and called on EU members to restrict the groups' activity in a non-binding resolution. Also last week the House Committee on International Relations, marked up Resolution 101, noting Hezbollah's capabilities are "equal to ... if not far more capable of" Al-Qa'ida. The Resolution also noted, "Whereas leaders of Hezbollah have made statements denouncing any distinction between its 'political and military' operations, such as Hezbollah's representative in the Lebanese Parliament, Mohammad Raad, who stated in 2001 that 'Hezbollah is a military resistance party .... There is no separation between politics and resistance.'"

Many pro-Syrian Lebanese politicians have in fact discussed both support of Hezbollah as well as how the terror organization is an integral part of the government given that over 10% of the parliament is made up of its members. For example, the international affairs advisor for Lebanese President Lahoud, Dr. George Dib appeared on Al-Manar TV on January 9, and explained Hezbollah is "the only card Lebanon can play" and that the government needs to "keep the resistance safe."

Lebanon's Defense Minister Abdul-Rahim Murad gave an interview to Egypt's Dream TV on November 28 in which he discussed Hezbollah's freedom to attack Israel, and how the terror organization receives its weapons from "Iranian ... Syrian and Lebanese aid." He also discussed relations between Lebanon's army and Hezbollah, "there is complete coordination."

The anti-Syrian rallies in Lebanon over the past few weeks was made up of anti-Hezbollah demonstrators. They included many Lebanese Christians, as its main newspaper Al-Nahar has been particularly vocal on this issue. As far back as August 14, 2003, its editor wrote, "Our controversy with Hezbollah centers on the fact that it has no right to exist as an armed movement within the state ... to carry out its own policies as if there were no government institutions." Another editorial on May 30, 2003 called on "disarming" Hezbollah and singled out Sheikh Nasrallah, "Nasrallah is urging [people to launch] an Iraqi Intifada against the American presence in Iraq, to continue the Palestinian Intifada, and to continue the armed resistance in southern Lebanon ... Who authorized Nasrallah to represent all the Lebanese, [and] to make decisions for them ... Did Nasrallah appoint himself secretary general of all the Lebanese and the whole Arab world?"

Nasarallah is known for his fiery speeches to crowds screaming "Death to America" and "Death to Israel." At last week's pro-Syrian rally, he declared, "To Syria we say long live Assad's Syria." He also made a direct threat to the U.S., "I address the following to America... I say to President Bush... And to Ms. Condoleezza Rice... You are wrong about Lebanon." He asked the crowed, "Are you Lebanese afraid of the American naval fleets? These naval fleets have come in the past, and were defeated, and if they come again, they will be defeated again."

Calling the U.S. "the Great Satan" last month during another rally, Nasrallah stated, "We consider the current administration an enemy of our [Islamic] nation ... Because it is the greatest plunderer of our treasures, our oil, and our resources ... This American administration is an enemy. Our motto, which we are not afraid to repeat year after year, is: 'Death to America.'"

As the events of the past few weeks took place, many pundits predicated the emergence of Hezbollah as the main power of Lebanon. This theory could turn out to be pre-mature as the massive Hezbollah sponsored rally was followed by an even larger anti-Syrian rally. The real message sent by those in attendance, including many Muslims, was directed not only to Syria but to Hezbollah.

The tipping point for which faction emerges in control of Lebanon will likely come down to how much pressure the West ends up placing on Syria.

Turkey's Spiritual Submission

March 23, 2005

A picture is worth a thousand words. A couple months after Turkey's prime minister, Recep Tayyip Erdogan, was elected to office representing the Justice and Development Party, the Turkish daily Star Gazette ran a photo on July 10, 2003, that shows Afghan jihad leader (and Taliban and Al-Qaeda ally) Gulbuddin Hikmatyar sitting with two men kneeling at his feet. The man on the right is Mr. Erdogan. The caption reads, "Taliban in the armchair, kneeling is the Prime Minister." It is important to recognize the significance of sitting at one's feet in Islamic tradition: It implies spiritual submission.

Mr. Hikmatyar has long-established ties with Osama bin Laden and is responsible for offering to shelter him in Afghanistan after he fled Sudan in 1996. Following September 11, 2001, Mr. Hikmatyar pledged allegiance with the spiritual leader of the Taliban, Mullah Omar, to launch a guerrilla war on the Afghan government and American troops there. Mr. Hikmatyar was named in Executive Order 13224 as a "Specially Designated Global Terrorist" during the same month that Mr. Erdogan was elected.

After Mr. Erdogan came to power, press reports in the West displayed some initial concern for a possible Islamist tilt by Turkey. As the Voice of America reported this week, the "underlying U.S.-Turkish tensions is the growing presence of Islam." The VOA went on to discuss how America had been in support of Turkey's experiment with a party that originated from the Islamist movement and that when it initially came to power it was viewed as something that would prove that Islam and democracy are compatible.

The VOA quoted a professor explaining why Turkey's opposition to American activity in Iraq does not justify anti-Americanism by members of the Turkish government and press corp. "When you have serious newspapers publishing articles about the U.S. having a secret weapon that makes earthquakes and that Istanbul is the next target ...When you have newspapers that publish all kinds of scurrilous articles about the U.S., that is more worrisome. The problem is that some Turkish politicians have joined the fray and have accused the U.S. of genocide and all kinds of other activities in Iraq," VOA recounted the professor saying.

Since the Islamists came to power, there has been a noted increase in anti-Americanism. A BBC poll that found that 82% of Turks are anti-American and one of Turkey's current best-sellers involves a fictional futuristic war with America. Additionally, Turkish protesters laid black floral wreaths outside the American embassy during Secretary of State Rice's visit last month.

MEMRI's Turkish Media Project monitors Turkey's press, and will soon cover TV channels, sermons from mosques, and textbooks. To date, this project has found rampant anti-Americanism. One example includes the Turkish Islamic daily Yeni Safak, known for unofficial ties to the current Turkish administration (such as the fact that the owner of the paper recently married a child of Turkey's Prime Minister).

Shortly after the American presidential election last year, Dr. Husnu Mahalli, a columnist at Yeni Safak wrote on November 17: "When will we understand that this is a new Crusader war?... When will we realize that American aggression targets ... all Muslim countries and peoples?... After bombing the mosques of Fallujah, the American soldiers desecrate them by urinating on and soiling their walls. After raiding homes, American soldiers strip the women and girls naked and molest them ... and this is why 59 million Americans voted for Bush."

The most recent casualty of the American-Turkish rift is American ambassador to Turkey, Eric Edelman, who has resigned his post. Some reports point to a statement Mr. Edelman made and was misquoted about dealing with Turkey's President Sezer's April visit to Syria as the reason. It is plausible to suggest issues surrounding the Iraq war are partially responsible for Turkey's increased anti-Americanism. But given that the country is ruled by an Islamist government whose president has given his spiritual devotion to an ally of Al-Qaeda, raises many questions which have yet to be explored.

Pope John Paul II's Visits to the Middle East

April 6, 2005

To mark the turn of the century, Pope John Paul II made a historic visit to the Middle East. In Israel he paid homage to the victims of the Holocaust by going to Yad Vashem and also visited the Wailing Wall in Jerusalem. At the time, these gestures enraged many Muslim leaders.

The pope also met with the grand mufti of Jerusalem, Ikrime Sabri, imam of the Al-Asqa mosque and the highest religious authority in Palestinian Islam. On the eve of the pope's visit to the Al-Asqa mosque, the grand mufti gave an extensive interview to the Italian newspaper La Republica, on March 24, 2000. The title of the article was "Too Many Lies About the Holocaust, Wojtyla Free Us From the Jews."

When asked his thoughts on the pope's visit to Yad Vashem the mufti answered: "Whether I approved of it or not, he would have gone anyway. The Jews have found a formidable way to gather solidarity from the world ... Well, the Crusades lasted 200 years, not five years like World War II, and many more Muslims than Jews were killed under Nazism. But nobody talks about the Crusades or compensation for us." The mufti also said, "Six million Jews dead? No way, they were much fewer. Let's stop with this fairytale ..."

The mufti was also asked if the pope would be allowed to pray in the Al-Aqsa mosque and explained, "It would not be a good idea ... such a gesture could be wrongfully interpreted by Christians in the future ..." In the interview he explained, "I will remind him [the pope] that according to Allah's will, the holy Muslim places belong to us only."

In February 2000, the pope also went to Egypt. His itinerary included a meeting with Sunni Islam's most prominent religious figure, the grand imam of Al-Azhar, Mohammad Sayyed Tantawi. In the past Mr. Tantawi has made statements in support of suicide bombings, killing Jews, as well as more recently support for jihad against American troops in Afghanistan and Iraq. During a press conference in Egypt, in what could be implied as a veiled reaction to such sentiment, the pope stated, "To do harm, to promote violence in the name of religion, is a terrible contradiction and a great offense."

In May 2001, the pope also went to Damascus and was greeted by Bashar Assad. The Syrian leader's welcome to the pope was an anti-Semitic diatribe: "They [the Jewish people] try to kill the principle of religions with the same mentality that they betrayed Jesus Christ and the same way they tried to betray and kill the prophet Mohammad." His Syrian itinerary included a visit to the historic Al-Umawi mosque.

From Mecca, Saudi Sheikh Adnan Ahmad Siyami responded to the pope's visit to Syria and his call for peace between religions: "The pope's recent visit to Syria, to the Al-Umawi mosque is, without a doubt, another manifestation of that call. The call by [the pope] - may Allah punish him as he deserves - to the people of the [different] religions ... is nothing more than an audacious call for the unification of religions. ... This pope, the head of the Catholic Church, and those behind him ... are the descendants of the Spanish inquisitors who tortured the Muslims most abominably. ... They are the descendants of those who led the Crusades to the Islamic East, in which thousands of Muslims were killed ... They are the perpetrators of the massacres in Bosnia-Herzegovina ... in Kosovo, in Indonesia, and in Chechnya ... Can we expect compassion from these murderous wolves?"

The Saudi sheikh added, "There can be neither an agreement nor a meeting point between the people of Islam and the Jewish and Christian People of the Book. ... How can we allow the Catholic pope's talk of a need to find meeting points and agreement between Islam and Christianity... Is it conceivable that there should be agreement and a meeting point with those who fabricate terrible falsehoods about Allah ... claiming that Jesus, peace be upon him, is his son?!"

Many Islamist Web sites have eulogized the pope with messages critical of Arab satellite television for highlighting John Paul II's call for peace and dialogue between all religions.

Next week's column will discuss the challenges the next pope will face in regard to the Muslim world.

Islam and the Vatican

April 13, 2005

Following Pope John Paul II's visits to the Middle East in 2000 and 2001, some prominent Muslim leaders openly discussed the future dominance of Islam in Europe, including a conquest of the Vatican.

As the identity of the next pope is decided, one of the pressing issues awaiting him is the growth of the Muslim community in Europe, part of which has Islamist inclinations. As the New York Times reported yesterday, one of the most important issues the next pope will face is the "increasing secularism in Europe, contrasting with the religious revival in the Islamic world ... and the rising number of Muslim immigrants in Europe."

Sheikh Muhammad bin Abd Al-Rahman Al-'Arifi, a Saudi and imam of the mosque of King Fahd Defense Academy, predicted Muslims would gain control of the Vatican. Citing a Hadith posted on the Kalemat Web site in 2002, he said, "We will control the land of the Vatican; we will control Rome and introduce Islam in it. Yes, the Christians, who carve crosses on the breasts of the Muslims ... will yet pay us the Jiziya [poll tax paid by non-Muslims under Muslim rule], in humiliation, or they will convert to Islam ..."

Sheikh Yousef Al-Qaradhawi, a spiritual leader of the Muslim Brotherhood who is head of the European Council for Fatwa and Research and the founder of the International Council of Muslim Scholars, posted a fatwa, or decree, on the Web site www.islamonline.net, in 2002 about the "signs of the victory of Islam" in Europe. Also, citing a well-known Hadith, Al-Qaradhawi wrote, "The Prophet Muhammad was asked: 'What city will be conquered first, Constantinople or Romiyya?' He answered: 'The city of Hirqil [i.e. the Byzantine emperor Heraclius] will be conquered first' - that is, Constantinople ... Romiyya is the city called today 'Rome,' the capital of Italy ... and we hope and believe [that it too will be conquered]."

Al-Qaradhawi elaborated on what this means today. "This means that Islam will return to Europe as a conqueror and victor, after being expelled from it twice ... I maintain that the conquest this time will not be by the sword but by preaching and ideology ..."

On his weekly religious program on Al-Jazeera in 1999, Al-Qaradhawi made similar statements: "All right, Constantinople was conquered, and the second part of the prophecy remains, that is, the conquest of Rome. This means that Islam will return to Europe. Islam entered Europe twice and left it ... Perhaps the next conquest ... will be by means of preaching and ideology. The conquest need not necessarily be by the sword ... Perhaps we will conquer these lands without armies. We want an army of preachers and teachers who will present Islam in all languages."

The following year, Al-Qaradhawi elaborated on the importance of European Da'wa (spreading Islam to non-Muslims): "Europe will see that it suffers from materialistic culture, and will seek an alternative ... Islam will return to Europe and the Europeans will convert to Islam."

Other Muslim figures who have discussed an Islamic conquest of the Vatican include the Palestinian Authority's Deputy Minister of Awqaf, Sheikh Yousef Juma'a Salameh; Saudi Sheikh Naser Muhammad Al-Naser, and Sudanese Sheikh Muhammad Abd Al-Karim.

Meanwhile, the population of Muslims in Europe has grown exponentially and as Al-Qaradhawi predicted, that has happened not by the sword but through preaching. In May 2000, there were reportedly about 1 million Muslims in Italy, a number that doubled in just 10 years, and more than 10,000 Italian-born converts, with the number of mosques and Islamic cultural centers having gone from 12 to 400 in the past 16 years.

Since September 11, 2001, European governments have started to monitor imams. After last year's Madrid bombings, the Spanish government proposed monitoring the sermons of imams. Following the murder of a filmmaker, Theo Van Gogh, the Netherlands started a program to train imams and no longer allows other imams from outside its borders. In France, where 10% of the population is Muslim, the government has started to expel Islamist imams. In Britain, Islamist imams are now on trial for connections to Al-Qaeda. Italy has also expelled imams for pledging allegiance to Osama bin Laden.

The issue of Islam in Europe can certainly be expected to be high on the agenda of the next pope.

Blueprint for Reform in the Mideast

April 20, 2005

While the buzz in the Arab world has been about the latest United Nations Development Report, a small conference in the region also quietly made headlines last week.

The 16th Arab National Congress was held in Algiers from April 6-9, with participants from throughout the Arab world. The focus of the four-day conference was mainly Iraq, Sudan, Lebanon, Syria, and the Palestinians.

The Congress produced a document highlighting the main points discussed, most of which centered around a "U.S.-Zionist project" targeting the Arab nation. It called for supporting "resistance and continued struggle by all means available." Arab regimes that "flock to appease" the West were denounced: "The ANC reaffirms that such appeasement and conciliation will only drive our enemy to more arrogance and sense of superiority, increasing his conviction that force is the only means of dealing with Arabs."

The ANC document devoted five points to Lebanon and Syria that discussed thwarting the "U.S.-Zionist assault" on those countries and Hezbollah was praised for its resistance activity. Four points of the document were devoted to Sudan and discussed the "U.S.-Israeli imperialist scheme against the Arab nation," while Arab television networks were called upon to uncover the truth behind these schemes.

Nine points were devoted to Iraq - in particular supporting anti-American activity there. The ANC called for enhancing the actions of the resistance and for liberating the country. The Iraqi people fighting "the occupation" were praised, and the document promoted the creation of a national front to liberate Iraq.

The ANC recommended that Arabs throughout the world join forces with the resistance and participate in the fight: "[We] call upon popular political forces all over the Arab world to activate the Arab street's support for the resistance; affirms the right of everyone in the nation to join forces with the resistance, considering its support by all its Arab neighbors, by all means possible, a duty of top priority; calls upon the efficiencies of the Arab popular Council to Support Iraq to continue their activity, strengthen these efficiencies, providing media support to Iraqi resistance."

The "Palestine question" was the subject of eight points. They included a call for the continuation of the intifada. America and Europe were "condemned" for placing Hamas and Islamic Jihad on lists of terrorist organizations.

The ANC encouraged the Arab world to financially support the intifada and for Arab states with normalized relations with Israel to reconsider their positions. The ANC also warned of falling into the "U.S. and Zionist trap aimed at breaking psychological barriers in this existential conflict through contacts, visits, or shaking hands with enemy officials." Palestinian "militancy" was praised, as was Congress member and Palestinian Bishop Attallah Hanna, for taking "militant" positions regarding confronting Zionist actions.

Bishop Atallah has been on the record numerous times supporting suicide attacks. On January 17, 2003, in the square of the Church of the Holy Sepulcher in Jerusalem, he led a protest against "American threats to attack Iraq" and expressed support for terrorism. According to Al-Quds Al-Arabi, Bishop Atallah stated, "We declare publicly our blessing, support, and legitimization of the brave Palestinian resistance [carried out] by any means, including the brave Fidaiyin martyrdom operations."

He added: "[The martyrs] will be inscribed in the history of our Palestinian and Arab people in holy white letters. The voices of those who defame these acts of heroism and honor are nothing more than anomalous voices that do not represent Arab and Palestinian public opinion ... The Fidaiyin martyrdom operations aimed deep inside Israel are unique pioneer operations that deter the Zionist enemy ... This is Arab Palestinian land, that has no connection whatsoever to the Jews and the Zionists ... Martyrdom operations are an excellent and good way to resist the Zionist invasion of the Palestinian land. In front of the Church of the Holy Sepulcher, we bless the souls of the heroic Shahids and the families of the Shahids."

While the ANC was anti-American in nature, the subject of reform was present throughout the four-day event. Just as Lebanese opposition figure Walid Jumblatt grudgingly admitted last month that he was wrong about President Bush and his reform agenda, even the most anti-American conference in the Arab world is now following the blueprint laid out by the president. Surely, this is another sign of victory for the policy of democratization and reform now springing up in the Middle East.

Top 10 Conspiracy Theories: Part II

April 26, 2005

It's only April, and already 2005 has seen its share of conspiracy theories from the Middle East. They began to appear in the early days of January, immediately following the tsunami, which was blamed on the usual suspects: America and Israel. The most popular accused America, Israel, and India of performing nuclear tests in the Indian Ocean, which caused the disaster. The following are the top 10 conspiracies so far this year:

10) On Iran's Sahar TV, an Iranian former Education Ministry official spoke on a February 8 program devoted to a Zionist ship sailing around the world kidnapping children to harvest their organs.

9) The Palestinian Authority's ambassador in Sri Lanka, Attallah Quiba, was quoted by the Malaysian National News Agency on March 19 as stating that Palestinian leader Yasser Arafat was killed by Israelis using powerful lasers.

8) Egypt's adviser to the United Nations, Shahida Al-Baz, was interviewed on Iqra TV on February 10, discussing drug abuse in his country explaining why Israel is responsible: "Although some people say this is a conspiracy theory, it is not... There is an enemy. Israel serves as a channel for drugs. Israel profits from this financially, and at the same time corrupts our youth."

7) The Syrian daily Teshreen on March 7 claimed the CIA and Mossad were behind the attacks of September 11, 2001: "The September 11, 2001 events in New York and Washington were a pretext to justify the implementation of the new domination strategy, knowing that the real perpetrators of the tragic events have been a toughly-controversial issue as many analysts still argue that those events were the creation of the U.S. CIA and the Israeli Mossad."

6) The Saudi Royal family Web site Ain-Al-Yaqeen reported on April 8, "A Palestinian official has unveiled news about the Israeli forces use of killer radiation in its war against the Palestinians. The Director of Al Shifa medical complex in Gaza Jom'a Al Saqa said Israeli troops force the Palestinian crossing the borders from Egypt and Jordan to enter into close chambers, remove their clothes and subject them to electromagnetic radiations. He added that electromagnetic radiations cause dangerous... and these include damage in the blood cells, breaking of blood platelets, cancerous tumors, blindness, mental retardation, and infertility."

5) Saudi journalist Suheila Hammad appeared on Saudi TV Channel 1 on February 3 and discussed conspiracy theories: "I'm not saying I believe in conspiracy theories, but, by Allah, this is a conspiracy. [The Americans] say so. The neocons have said: 'Within 30 years, we will eliminate Islam.' Among the targeted regimes are Saudi Arabia, Egypt, Syria, and Iran. This is part of the existing conspiracy. It will divide all the Arab countries into ministates... Sudan is being divided so that Darfour will become a secular state... the south will become a Christian state, and there is a conspiracy and a secret agreement among the Egyptian Copts in the U.S. for establishing a Coptic autonomy in the south of Egypt."

4) A report in the Tehran Times of March 8, citing Al-Jazeera as its source, claimed that the CIA and the Mossad were involved in kidnappings and beheadings in Iraq as well as claiming the American military tried to assassinate Italian journalist Giuliana Sgrena in Iraq.

3) Following up from leading Arabic dailies in Saudi Arabia and Syria reporting that American soldiers are stealing the body parts of Iraqis, the general commander of the Iranian Revolutionary Guard Corps, Yahya Rahim-Safavy, told Iranian Channel 4 on February 6 that American soldiers at Fallujah harvested and sold the organs of dead Iraqis.

2) Discussing the killing of Lebanon's Prime Minister Hariri, Syrian columnist Hassan M. Yousef wrote in the Teshreen daily on February 16 that "the U.S. and Zionist entity" were behind it. A Tehran Times headline report on February 16 was titled "Mossad Behind Hariri Assasination." Rakan Majali contended in the Jordanian daily Al-Dustour on March 7 "that the killing ... was part of the American scheme to start a confrontation between Syria and Lebanon. "The latest hit song of Egypt's most popular singer, Abdel-Rahim, whose songs

include, "I Hate Israel" and other songs that blame the attacks of September 11, 2001, on America, holds Ariel Sharon accountable for Hariri's death.

1) Egyptian journalist Sayyed Nassar appeared on Dream 2 TV on February 6 to explain that the Iraqi elections were a theatrical production conducted by an American filmmaker.

A Brotherhood of Terrorists

May 4, 2005

In recent months, there have been multiple reports in Washington that America may begin a dialogue with the Muslim Brotherhood - the world's most prominent Islamist organization. As reported in U.S. News & World Report's April 25 issue, American intelligence officials have been meeting with members of the Brotherhood, and an unidentified source called them "probably more a part of the solution than a part of the problem." Some Washington think tanks have also begun to float this idea in the nation's capital.

It should be noted that the philosophy of the Brotherhood - which has inspired the likes of Al-Qaeda and Hamas (which defines itself as the group's Palestinian branch) - today is led by many telegenic spokesmen who often appear on Arab TV stations in favor of terrorist attacks. For example, Dr. Abd Al-Muni'm Abu Al-Futuh of the General Guidance Council of the Muslim Brotherhood appeared on Al-Jazeera TV on April 25, 2004, where he said, "We, as Muslim Brotherhood and as the world leadership of the Muslim Brotherhood, support the armed resistance in Iraq against the American occupier, 100%. I personally, as all members of the Muslim Brotherhood in Egypt and around the world, wish to carry weapons against the American occupation ... This is a matter of religion, not politics. When a Muslim land is occupied, Jihad becomes an individual duty ..."

The deputy leader of the Muslim Brotherhood, Mahmoud Al-Sayyid Ahmad Al-Habib, made a similar appearance on Hezbollah's Al-Manar TV, on April 8, 2004, and said, "The truth is that the resistance, whether in Iraq or in Palestine ... defends the nation's honor ... Therefore, the issue ... martyrdom (i.e., suicide) operations carried out by boys and girls, and also the operations carried out by the Iraqi resistance - these redeem self-confidence and hope, because a nation that does not excel at the industry of death does not deserve life."

Last year, Muhamad Madi Othman 'Akef took over as the Brotherhood's "supreme guide." At the time, he gave a series of interviews laying out the ideology of his group. He explained it was "a religious duty" to attack American and Israeli soldiers and, speaking to the Egyptian weekly Al-Arabi on January 18, 2004, he denied being in dialogue with America, which he referred to as "Satan."

Discussing the coming collapse of America, Mr.'Akef gave an interview to the Egyptian weekly Nahdhat Misr on January 21, 2004. He said, "The matter of America's attempt to take over the world is not new ... I expect America to collapse soon. The elements of this collapse in America [already] exist, and Allah is the Savior."

According to the Muslim Brotherhood Web site, Mr. 'Akef last year discussed the coming conversion of Europe and America to Islam. "I have complete faith that Islam will invade Europe and America, because Islam has logic and a mission. The Europeans and the Americans will come into the bosom of Islam out of conviction."

Having explained the pro-Jihad terrorist ideology of the Brotherhood, in this limited space it is difficult to understand how anyone in America would consider a dialogue with the group. In fact, this has infuriated many in the Middle East, including some of the most prominent pro-American reformists, who America should be listening to. For example, a Kuwaiti intellectual, Ahmad Baghdadi, wrote a series of articles in January 2005 in the Kuwaiti daily Al-Siyassa and Egyptian weekly Akhbar Al-Yaum calling for the outlawing and disbanding of the Brotherhood.

Calling the group a "terrorist organization," he questioned the "U.S.'s silence regarding the Muslim Brotherhood, despite the fact that [this organization] supports terrorism, whether by supporting bin Laden ... or by publishing terrorist fatwa's ... calling to destroy democracy ..."

Mr. Baghdadi warned America not to believe the Brotherhood and that there is the possibility it could moderate. "The policy of bribing religious organizations is a total failure, since these religious organizations are like snakes," he wrote. "You can never know when and where they will bite you. Thus, the best policy for dealing with [these] snakes is to kill them right off. ... Reliance on these organizations means the destruction not only of democracy, but also of the future, which the U.S. aspires to achieve through its 'Greater Middle East' program. Staying silent in the face of the Muslim Brotherhood's ... activities [only] multiplies its strength and reach. The day will come when the only solution for this will be confrontation, and it is better that this be now and not later."

It should be noted that Mr. Baghdadi's ongoing criticism of Islamists has led him to seek political asylum in the West.

Making Arab TV 'Jihad-Proof'

May 11, 2005

MEMRI's TV Monitor Project, which has now spent a year tracking Arab and Iranian satellite TV, has shown that as these channels are beamed into the West - and America in particular - they have the potential to incite viewers to jihad. France has recognized this danger and within the last year banned two channels: Hezbollah's Al-Manar TV and Iran's Sahar TV. The European Union as a whole is also looking to ban certain channels.

Arabic and Farsi speakers living in the West who depend on Middle Eastern satellite TV for news are inundated with daily messages of hate. In North America, the following are just some of the many satellite channels available: Al Jazeera, Al-Majd, Iqraa, ART, LBC, Nile, Future, MBC, NBN, Ajaman, Al Alam (Iran), Al Manar, and Orbit. State TV channels are available from Egypt, Saudi Arabia, Jordan, Iraq, Kuwait, Bahrain, Morocco, Sudan, Dubai, Tunisia, and Libya.

Middle East TV available by satellite also broadcasts leading Islamist imams espousing hatred. A spiritual leader of the Muslim Brotherhood movement and head of the European Council for Fatwa, Sheikh Yousef Al-Qaradhawi, for example, has a weekly show on Al Jazeera.

Claiming that 23,000 people in America converted to Islam in the month following September 11, 2001, Egyptian Sheikh Mahmoud al-Masari spoke on Al-Majd TV on March 2, stating "how beautiful" it was to see the World Trade Center buildings fall. The studio audience laughed out loud when he described how "wretched" Americans were punished that day.

Discussions by Muslims in the West calling for jihad are also frequently aired. The imam of London's Central Mosque, Sheikh Said Radhwan, called for jihad while speaking on Iran TV in August: "The Arab and Islamic nation must mobilize everything. Islamic law teaches us that when the people of a country cannot defend themselves - jihad is a personal duty that applies to all Muslims. We must wage jihad with our property and our souls."

Another Londoner, Muhammad al-Musawi, spoke to the same channel. In that October interview, he said, "What I'm saying is that a psychological war is being waged by hostile circles. They are trying to present themselves as victorious... You should analyze their reality: They are afraid and not vice versa. The Muslims should know that we are the stronger ones. We are stronger than them in a number of ways ... We are stronger psychologically because we trust in the true religion ... We have a mental strength that these enemies lack. They are internally defeated."

Sermons and interviews calling for killing Christians, Jews, and American soldiers, as well as confrontations with the West by Muslim religious figures, are aired by many of the satellite channels. Egypt's Sheikh Muhammad al-Mussayer of Al-Azhar spoke on Iqraa TV last July, saying that Islam should not merely be defended but that "Muslims today should confront."

Saudi Sheikh 'Aed al-Qarni also stated on Iqraa TV in December: "I pray to Allah ... that He will destroy the Jews and their helpers from among the Christians and the communists, and that He will turn them into the Muslims' spoils. I praise the jihad, the sacrifice, and the resistance against the occupiers in Iraq. We curse them all of them every night and pray that Allah will annihilate them, tear them apart, and grant us victory over them. The jihad in Fallujah is a source of pride ... Who can say even one word against this true jihad against these colonialist occupiers?"

The influence of Middle East TV satellite channels has been demonstrated repeatedly by the Iraqi confession show that appears on al-Iraqiya. Captured terrorists have stated time and again that they were influenced by the "sermons of Qaradhawi," "fatwas by leading Saudi religious authorities," and "watching Al Jazeera."

Policy-makers, especially those involved in fighting the war on terror, should recognize that within American borders, satellite television beamed live from the Middle East includes shows that call for the killing of Americans. This is one aspect of the war on terror that should be addressed and can easily be taken care of by blocking such broadcasts.

Next week, this column will focus on the potential of Arab and Iranian TV to incite Muslims in the West to anti-Semitism.

Banning Arab TV In America

May 18, 2005

Rep. Steven Rothman, a Democrat of New Jersey, and Rep. Joe Knollenberg, a Republican of Michigan, yesterday sponsored a MEMRI briefing on Arab and Iranian satellite channels available in America. Their broadcasts have the potential to incite viewers to jihad and contain messages of hate against Jews and Christians. As the content of such programming is exposed, it is hopeful that the Federal Communications Commission - which answers directly to Congress and the mission of which includes regulating international television in America - would consider legal proceedings against American companies that air Arab and Iranian TV channels spewing hatred.

Programs on these channels include statements clearly directed at Americans. During an official Friday sermon on August 27 aired on Sudanese TV, for example, a preacher stated that Jews engineered the attacks of September 11, 2001. He cited a quote fabricated by the Nazis and inaccurately attributed it to Benjamin Franklin: "The malaria microbe known as the Jews, which the U.S. carries in its stomach, will kill it sooner or later." He added, "These Jews hasten America's death. The U.S. must be aware. We offer this advice via the TV channels."

Similar Nazi falsehoods have resurfaced of late on other Arab and Iranian TV stations available in America, such as Al-Jazeera and Al-Manar. Tuning into those channels can be strikingly similar to watching Nazi propaganda. Programming includes shows devoted to distortions in Jewish religious texts, "The Protocols of the Elders of Zion," Jewish corruption throughout history, and the blood libel. On a UAE/Saudi channel Al-Majd program of March 22, 2004, the secretary-general of the Islamic Action Front Party of Jordan, Sheikh Hamza Mansour described the "sick psychological nature" of the Jews, adding that they make "matzoth out of innocent children's blood."

In August on Saudi Arabia's MBC TV, Saudi Sheikh Abd al-Qader Hammad said, "From the day civil strife began in Islam, the Jews were behind it. There is no evil in the world that the Jews are not behind. Search in the depths of history ... and you will find the finger of the Jews behind it."

"Jewish treachery is the nature of this race. It runs in their blood and passes from generation to generation," 'Adnan abu Al-Qattam said in an official Friday sermon that aired on Bahrain TV July 30. "The Jews proved to history ... they are a corrupting people ... History clearly attests to their... despicable behavior," he said. He then cited a Koranic verse calling Muslims "to kill Jews."

Saudi professor Walid al-Shuzi explained that according to Islam and the Koran, Muslims "have a duty to hate Jews and Christians," during an Al-Majd TV broadcast of March 18.

Mr. al-Shuzi's topic? "So-called" anti-Semitism.

"The Protocols of the Elders of Zion" is often cited as proof of the Jews' evil nature. A Danish cleric, Nabel Shaker Al-Taleqani, appeared on Iran's Al-'Alam TV on October 30. While speaking of Jewish involvement in terrorism, he said, "Go back to the book 'The Protocols of the Elders of Zion.' We must return to the 24 protocols, to these secret documents." The secretary-general of the Islamic Universities' Association, Dr. Ja'far Abd Al-Salim, cited "The Protocols of the Elders of Zion" on Al-Manar TV on April 22, 2004. Mr. Al-Salim said it proved that Zionist-Jews "main goal is to humiliate the human race." He also said that only Jewish companies make drugs and pornography. Egypt's Al-Ahzar Sheikh Muhammad Al-Mussayar appeared on Iqraa TV on July 12 to speak about a global campaign against Islam and the need to "confront our enemies ... expose their secret beliefs" - which he said are laid out in "The Protocols of the Elders of Zion."

Arab and Iranian TV also leap to prove the vileness of the Jews through distortions of religious texts such as Talmud and also the Torah. Prominent Muslim religious figures are not above taking this tack. For example, the leading Palestinian Authority religious figure, Mufti Ikrima Sabri, appeared on Al-Majd TV on February 20. During the program he said, "Anyone who studies 'The Protocols of the Elders of Zion' and specifically the Talmud will discover that one of the goals of these protocols is to cause confusion of the world."

These examples are but a few of the dozens that appear on satellite TV beamed daily into America. There is no doubt that such programming ought not be accessible in America.

Sermons As Political Propaganda

May 25, 2005

Under Yasser Arafat, Palestinian Authority sermons were filled with incitement, including calls to kill Jews and destroy America. Few in the West at the time understood the danger posed by these religious proclamations; those intimately involved in the peace process were often willing to overlook them.

The Khatibs, or preachers, who give such sermons, are paid employees of the PA. Their messages are broadcasted live every Friday at noon from mosques controlled by the PA and appear on official PA TV.

Sheikh Ibrahim Mudeiris was a favorite Khatib of Arafat. Mr. Mudeiris is based at the Sheikh 'Ijlin Mosque in Gaza and makes regular appearances during the coveted Friday noon timeslot. One of his notable proclamations about America, from September 5, 2003, was, "America will collapse ... we consider America to be our no. 1." On May 21, 2004, Mr. Mudeiris made reference to America again: "the American dog, the Byzantine dog of our days ... To the Byzantine dog, you son of a bitch, I have brought an enormous army upon you, that starts here and ends here." In sermon from March 21, 2003, he said, "Allah will drown the little Pharaoh, the dwarf, the Pharaoh of all times, of our time, the American president. Allah will drown America in our seas, in our skies, in our land ... America will be destroyed."

When Abu Mazen came to power following Arafat's death, he promised to end incitement. On December 3, 2004, he appeared live on PA TV at the President's Mosque as an audience member of an exceptionally moderate sermon calling for Palestinians to recognize and tolerate others, and to avoid extremism. Mr. Mudeiris was notably absent at the President's Mosque, but quietly reclaimed his pulpit and TV spot soon afterward.

Recent sermons by Mr. Mudeiris include one from April 15, 2005, in which he said that Muslims prisoners in Iraq and Afghanistan were forced to convert to Christianity. In his sermon on February 4, 2005, he called for Palestinians to conquer Israel. This year's New Year sermon was devoted to the destruction of America: "America has reached the top and we admit it, but it is headed for a bottomless pit, Allah willing. America, which is being led by its current president, to a bottomless pit. He leads it to death and destruction, Allah willing. America's grave was dug by Bush the day he invaded Afghanistan. Bush prepared this grave on the day he invaded Iraq."

Throughout the year leading up to Arafat's death, Mr. Mudeiris kept busy inciting Palestinians. In reference to Jews and Israel on November 5, he blamed "these apes and pigs" for Arafat's death. On March 12, 2004, he explained "the Jews today ... are avenging their ancient forefathers, the sons of apes and pigs ... They deserve death ... We strike more fear into their hearts than their Maker."

Mr. Mudeiris's sermon from May 13, 2005, claimed Muslims will rule America and called Jews a virus resembling AIDS. With a visit to the White House this week and following criticism of that speech, the PA's information minister, Nabil Shaath, who is in charge of PA TV, promised to "suspend Mudeiris and prevent him from delivering further sermons."

The call for his removal is a positive development, yet it should be taken in context. Mr. Mudeiris has been giving sermons for more than a decade, even during the period of the Oslo Accords when Mr. Shaath was in power. The PA's denouncement of a particular sermon comes not over a fundamental disagreement with Mr. Mudeiris's beliefs, but rather as a prelude to Abu Mazen's upcoming Washington visit. As one press report stated, "A senior PA official expressed fear that the publication of the contents of the sermon would reflect negatively on Mazen's upcoming talks with the U.S. president. 'This is the last thing we need now,' he remarked. 'This preacher has done grave damage to our cause.'"

Unlike American officials involved in the peace process under the Clinton administration, the Bush administration has been clear in its demands that Arab incitement end. When Abu Mazen visits the White House this week, he should be asked why an official religious representative of the PA has not only been inciting Palestinians against Jews and Israel, but also calling for the destruction of America.

The Problem With Turkey

June 1, 2005

In January 2005, MEMRI launched its Turkish Media Project, which translates press reports from that country. The need for the project arose following reports of unprecedented anti-Americanism and anti-Semitism, beginning in November 2002 when the Justice and Development Party, under the leadership of its Islamist prime minister, Recep Tayyip Erdogan, took over the country.

Certain corners called on MEMRI to stop its work on this subject following the project's first few releases: Turkish lobbyists in Washington, D.C., the foreign ministry of Turkey, and leading American Jewish organizations have all attempted to stifle the research. (This calculated attempt will be the subject of a column in the coming weeks).

Additionally, on April 6, 2005, the vice consul at the Turkish Consul General, Murat Omeroglu, sent an animated letter to The New York Sun in response to this author's March 23, 2005, column, "Turkey's Spiritual Submission." Mr. Omeroglu argued that the column falsely accused Turkey of being anti-American and failing to participate as a full partner in the war on terror. The article focused on neither point, although anti-Americanism in Turkey is rampant - a BBC poll conducted last year found Turkey to hold the most anti-American sentiment in the world.

"Turkey's Spiritual Submission" rather spoke to the cause of the recent tension between America and Turkey - the Islamist Justice and Development Party and its leader. It included a photo, which originally appeared in the Turkish daily Star Gazette, that showed an Afghan jihad leader with ties to the Taliban and Al-Qaeda, Gulbuddin Hikmatyar, with two men kneeling at his feet. One was Mr. Erdogan. The Star Gazette's caption read, "Taliban in the armchair, kneeling is the Prime Minister." My column highlighted the cultural significance of sitting at another's feet in Islamic tradition: It implies spiritual submission. Mr. Hikmatyar has a long-established relationship with Osama bin Laden and offered to shelter him in Afghanistan when he fled Sudan in 1996. Following September 11, 2001, with the spiritual leader of the Taliban, Mullah Omar, Mr. Hikmatyar pledged to launch a guerrilla war on the Afghan government and American troops there. He was named in Executive Order 13224 as a "Specially Designated Global Terrorist" in the same month that Mr. Erdogan was elected.

What does the fact that Turkey's prime minister remains so close to an enemy of America say about the country's government as a whole?

It is helpful to have additional background information on Mr. Erdogan to understand the present strained relationship between America and Turkey. Mr. Erdogan was active in the now-defunct Islamist National Salvation Party until a 1980 military coup disbanded the party. When its former members founded the Islamist Welfare Party, he became chairman of the party in Istanbul. By 1994 the party emerged as Turkey's largest - just as Mr. Erdogan became mayor of Istanbul. Two years later, the Welfare Party was declared unconstitutional and was disbanded on the grounds that it threatened Turkey's secularity.

On December 12, 1997, at a public meeting in southeastern Turkey, Mr. Erdogan read a passage by renowned Turkish poet Ziya Gokalp: "The mosques are our barracks, the domes our helmets, the minarets our bayonets, and the faithful our soldiers." As the BBC reported, "his pro-Islamist sympathies earned him a conviction in 1998 for inciting religious hatred."

Following Mr. Erdogan's trial and conviction, he was sentenced to a short jail term, which barred him temporarily from politics. (The Turkish Parliament changed the constitution in 2001, allowing those who had served time in jail to serve in public office). By 1999, Mr. Erdogan's Welfare Party merged with the Virtue Party - which was also banned. This led to the creation of the Justice and Development Party that he now heads.

With Mr. Erdogan scheduled to visit to Washington, D.C., next month, it should be noted that Turkey has developed a PR machine that is working overtime to create the impression that the ruling party is an ally of America and a friend of the West. Mr. Erdogan's personal history should not be overlooked in assessing his political and religious inclinations.

A True Moderate For Iran?

June 8, 2005

On June 17, Iran will hold presidential elections. The leading candidate to replace President Khatami is one of Iran's most influential politicians, Ali Akbar Hashemi Rafsanjani. Mr. Rafsanjani served two terms as Iran's president, from 1989 to 1997, and has been part of the conservative establishment since the Islamic Revolution of 1979. He is now the chairman of the powerful Expediency Council.

He is vocal in his opposition to terrorism in general and state-sponsored terrorist regimes in particular. According to him, the leading offender in the latter camp is America - not Iran. In an interview with Jomhouri-ye Eslami on July 12, 2003, Mr. Rafsanjani claimed that America created the Taliban and Al-Qaeda in Afghanistan to oppose Iran, but that they have now become "trouble" for Americans. On March 21, 2004, Gooya.com quoted him as saying that world public opinion rejects terrorism, "but the problem is that the U.S. is the standard-bearer of terrorism."

Three days after September 11, 2001, in an interview with the Islamic Republic News Agency, Mr. Rafsanjani blamed America for disseminating "accusations against the Muslims, inciting public opinion, and fanning the winds of war because it cannot find the main agents behind the crime." In an interview with Hayat-e on the one month anniversary of the attacks, he said, "If the Americans also dream of reaching Pakistan, Afghanistan, and Central Asia, and of containing Iran, there will be an extended Jihad within the Islamic world; there will be no security, and America will be dealing with new circumstances."

On American criticism of Iran's nuclear program, he told the IRNA on September 16, 2003, that the "U.S. campaign against Iran's peaceful nuclear activities is nothing except declaring war on Islam" and added that Iran has a right to acquire nuclear technology. "If we pursue our plans properly, the enemies would fail to prevent us from our scientific activities," he said. It should be noted that he personally signed an agreement with North Korea for the delivery of North Korean missiles in 1985 valued at $500 million, which assisted the development of Iran's Shihab missile.

Mr. Rafsanjani has followed in the footsteps of the late Ayatollah Khomeini in many ways - for example, marking World Al-Quds (Jerusalem) Day by supporting Palestinian Arabs. This holiday is celebrated on the last Friday of the month of Ramadan. During the official sermon on December 14, 2001, at Tehran University, Mr. Rafsanjani stated that Muslims must surround colonialism (i.e., America and Britian) and force the colonialists to see whether Israel is beneficial to them. If one day, he said, the world of Islam comes to possess the weapons (nuclear ones) currently in Israel's possession, this global arrogance would come to an end. The use of a nuclear bomb in Israel would leave nothing on the ground, he said.

Mr. Rafsanjani is influential in dictating Iran's policy toward Palestinian Arabs. He named his third son Yasser - after Arafat. He frequently calls for the destruction of Israel. In an interview with the IRNA on April 18, 2004, he said: "Undoubtedly, the blood of Palestinian martyrs ... would give rise to a new momentum among Muslims and sooner or later, the occupying Israeli government would be doomed to extinction."

In an interview with Kayhan on December 15, 2001, he said, "Jews should wait for a day when this 'extraneous matter' [Israel] would be removed from the region and from the world of Islam, and those who have gathered together in Israel would one day be dispersed again."

Following the assassination of the spiritual leader of Hamas, Mr. Rafsanjani said that no death could be more fitting for Sheikh Yassin, "who had spent his blessed life in a campaign against Israel, than to be martyred by the most merciless anti-freedom and inhumane criminals." He predicted that the "dastardly" killing would "expedite the Palestinians' ultimate victory over Israel."

Mr. Rafsanjani explained in an interview with Al-Wifaq on December 15, 2001, that the Palestinian jihad is on "the forefront of global Jihad and is the mother of many Islamic movements, ranging from the [1979] Islamic Revolution in Iran to Islamic movements in Lebanon, Afghanistan, Malaysia, Central Asia, and many other Muslim states. All these Islamic movements support the Palestinian Jihad."

Many in the West have labeled Mr. Rafsanjani a moderate. His stances on America, peace in the Middle East, and Iran's nuclear weapons speak for themselves.

Al-Zarqawi Conspiracy Theories

June 15, 2005

The Iraq War is replacing the attacks of September 11, 2001, as fodder for conspiracy theories emanating from the Middle East - in particular, questions surrounding Abu Mus'ab al-Zarqawi. Does he exist? Where is he hiding? Is he actually an American or Jordanian agent? Is he dead?

Print and television reports in the Arab and Iranian press regularly question Mr. al-Zarqawi's existence. A member of the former Iraqi Governing Council, Fadhl al-Rube'i, was interviewed on the Lebanese channel New TV, on May 16, 2004. He called Mr. al-Zarqawi mythical and a creation of the Pentagon's "disinformation" center.

The spokesman for radical Iraqi Shi'ite leader Muqtada al-Sadr, Sheikh Abd al-Hadi al-Daraji, likewise stated in an April 4 interview with the Israeli-Arab weekly Kul Al-Arab that Mr. al-Zarqawi was an "ambiguous, imaginary, and made-up figure," created to justify American operations in Iraq. According to a report in the Baghdad newspaper Al-Mu'tamar, on October 28, 2004, fliers distributed by Sunni clerics in Fallujah maintained, "there is no Zarqawi in existence and that such person was an American creation."

An international relations expert at Egypt's Al-Ahram Institute, Said al-Lawindi, appeared on Egyptian channel 1, on November 16, 2004, and alleged that Mr. al-Zarqawi was a fabrication: "Where is this al-Zarqawi? I have read even in the French press, in Le Monde, that anyone can show a photograph of someone who died decades ago and claim that it is al-Zarqawi. This is an attempt by the U.S. to emphasize ... you can't go on forever talking about bin Laden, who has also become a myth."

An article titled, "Zarqawi the Terror Monster: But Does He Really Exist?" appeared in the Saudi daily Arab News, on October 26, 2004, and stated, "Zarqawi has been built up into an almost legendary figure ... Zarqawi is suspected of direct involvement in the kidnap and beheading of several foreigners in Iraq ... But many question his very existence." The article also discussed the battle in Falluja: "The people of Falluja, however, insist that they have never seen the man or heard about him except through the media. So, where is Zarqawi or, indeed, does he exist?"

Over the past few years the Iranian press has blamed America and Israel for most of the terrorist attacks in Iraq, especially the ones that Mr. al-Zarqawi was believed to have masterminded. The same exact newspapers and TV programs have of late changed their story - they now point to Jordan as well.

In a December 26, 2004, Tehran Times article titled, "What Does Jordan Want From Iraq," Hassan Hanizadeh wrote: "After the overthrow of Saddam, officials of ... Jordan sent Abu Musab al-Zarqawi to Iraq to carry out destabilization activities ... Jordan is implementing a disinformation campaign against the Islamic Republic of Iran and the Iraqi Shias ... Jordan intends to reestablish the Ba'ath regime in Iraq and re-marginalize the Shias, in order to resolve the Palestine issue through the resettlement of five million Palestinian exiles in western Iraq."

At the end of May, amid reports that Mr. al-Zarqawi was wounded in battle, rumors circulated throughout the Middle East discussing whether he really did exist - was he alive? On June 1, 2005, Iran's Mehr News Agency, reported, "Al-Zarqawi has received treatment at a hospital in the Jordanian capital Amman amid tight security by Jordanian intelligence forces and under the supervision of the Jordanian King Abdullah II ... There is reliable news in Jordan that al-Zarqawi was transferred to the Al-Hussein clinic in Oman ... The news comes after reports previously claimed that al-Zarqawi was receiving medical treatment in Iran."

Responding to claims that he was in Iran, on May 31, 2005, in an article titled, "Is al-Zarqawi in Tehran?" Mr. Hanizadeh wrote in the Tehran Times that claims of Mr. al-Zarqawi receiving medical treatment in Iran was, "similar to an April's Fools Day joke."

The first week of June 2005 saw reports in Saudi dailies Al-Madinah and Okaz, as well as Western publications such as the Associated Press, that quoted insurgents attesting to Mr. al-Zarqawi's death. Al-Mandinah claimed he was buried in a Falluja cemetery and that he left a will instructing Osama bin Laden to appoint his successor.

As this column has repeatedly discussed over the past year, in the Arab world and Iran conspiracy theories surround almost every major news story. The subject of Mr. al-Zarqawi is just one of the latest.

Arab, Iranian Press Attack Bolton

June 22, 2005

Over the past few months the Arab and Iranian press have launched persistent attacks on President Bush's nominee for U.N. ambassador, John Bolton. These are the same outlets that blame America for terrorist activity throughout the world, claim American soldiers cannibalize Iraqis, and make racist comments about Secretary of State Rice.

Arab and Iranian criticism of Mr. Bolton ranges from a dislike of his physical appearance to his position on the war on terror to his role in defeating a U.N. resolution equating "Zionism With Racism" and his stance against Palestinian terrorist organizations.

An anti-American writer for the London Arabic daily Al Hayat, Jihad Al-Khazen, has devoted a series of columns to haranguing Mr. Bolton. Two were printed during the week of April 25. The first, "John Bolton: The Diplomat Or The Thug," called him "corrupt merchandise" and a "gangster." In the second, "The Problem Is His Politics," Mr. Khazen described Mr. Bolton as "a warmonger and a supporter of American unilateralism and in favor of Israel." Mr. Khazen continued by belittling Bolton's looks: "His appearance is inappropriate, he looks like a sea animal."

The Iranian press has likewise railed against Mr. Bolton. An April 27 article in the Tehran Times headlined, "Wrong Man at the Wrong Time" called his nomination "appalling" and said it "sent shivers" through the United Nations. On his physical features and his position against Iran's nuclear program, the article stated, "The mustachioed hawk, who wants to see his nightmare worldview translated into reality, has repeatedly claimed that Iran is seeking a nuclear weapons program ... But Bolton is actually the wrong man at the wrong time because he represents the intolerant and unilateralist tendency in U.S. foreign policy." A May 1 report from the Mehr News Agency said that Mr. Bolton's ties to "hardcore neo-conservatism" prove that he aims to stop "the establishment of a Palestinian state."

The Syrian press has been particularly nasty. In "Why Bolton?" which appeared in the government daily Tishreen on May 18, M. Al-Nahas criticized Mr. Bolton's alleged role in pressuring the State Department not to condemn the assassination of a Hamas leader, Sheikh Ahmed Yassin. Also in Tishreen, on March 27, R. Zein called the nomination a "deliberate and provocative appointment, which will further weaken the U.N. and make it eventually succumb to the hegemonistic dictate of the unilateral superpower." He added, "the arrogant superpower continues to run a tyrant course that has nothing to do with basic U.N. principles of right, justice, and equality, nor has it anything to do with peoples' national interests ... and arrogant U.S. policies of domination and hegemony, made it less safe and more dangerous."

Mr. Zein linked the nomination to "the new American-Zionist expansionist project targeting the ... Middle East and at serving new hegemonistic objectives drawn up in the global American strategy of all-out domination" and described Mr. Bolton's take on American policy in the region "sinister," "malicious," and "evil."

The Egyptian press has also stridently opposed Mr. Bolton's ambassador nomination. A March 24 article in the government-controlled Al-Ahram Weekly, "Red Alert," said President Bush "behaved like a terrorist" by nominating Mr. Bolton, and said that if he does get confirmed, Mr. Bush "will only add to his caricatured image and make a fool of himself."

In an Al-Ahram article called "American Nightmare," a political science professor at Cairo University, Hassan Nafaa, called Mr. Bolton and others in the Bush administration "the most virulent American ideologues." They were criticized for being pro-Israel and for being "firm believers in uncontested American global supremacy." Mr. Nafaa called Mr. Bolton a "slick attorney par excellence," and compared him to a "mafia don" for helping America "worm" its way through international law. He concluded with this: "Undoubtedly, Bush brought him in to slap the hands of anyone who has the audacity to remind America about international law ... But more importantly, his task is to ensure that anyone who falls out of line with American policy will not only incur Washington's wrath."

State-sponsored terrorist regimes such as Iran and Syria express open concern about Mr. Bolton's nomination - and moderate governments in the region, such as Egypt's, through its state press, also show fear that he will challenge their longstanding policies.

Saudis Seek Punishment Over Koran

June 29, 2005

Shortly following news reports alleging that copies of the Koran were desecrated at Guantanamo Bay, American officials at the highest level were quick to note that it is government policy to ensure that the Muslim holy books are treated with "utmost care" and "respect." During a White House press briefing on May 17, 2005, a spokesman, Scott McClellan, discussed how the allegations were being "exploited" and "used to incite violence" in the Muslim world.

In Saudi Arabia, members of the royal family, leading religious figures, and members of the state-controlled press spoke out on this issue. Some called for American soldiers to be tried by Shariah or Islamic law - implying a possible death sentence.

Saudi Arabia's Ministry of Foreign Affairs issued a press release that was posted on the Web site of the Saudi Embassy in Washington, D.C., on May 12. It stated: "The Kingdom urges the concerned United States authorities to launch a prompt investigation into these allegations. If these allegations proved true, it would be necessary to hold those responsible accountable for their actions and deter any repetition of such actions which are offensive to the Muslim community worldwide."

In a Friday sermon broadcasted live on May 20 on Saudi TV from Islam's holiest site, the Al-Haram Mosque in Mecca, Saudi Arabia's highest-ranking government-appointed cleric, Sheikh Abd Al-Rahman Sudais, responded to the allegations. "We are demanding that the transgressors who erred and performed this despicable act be interrogated and that the heaviest and the most preventive punishments be imposed on them," he said. It should be noted that Sheikh Al-Sudais planned to visit North America in May 2004, but following reports of MEMRI translated quotes from sermons in which he supported jihad against the West, and criticized America and Jews, Canadian members of Parliament initiated a debate that barred him from entering the country.

A report on May 22 in the Saudi Gazette elaborated on how some Saudis wanted to retaliate. The report stated that 18 Saudi scholars demanded that those involved in the alleged desecration of the Koran be tried by an Islamic court. The scholars also demanded the revival of a Saudi boycott of American products and for all Saudis detained at Guantanamo to be released. According to one report in Al-Jazeerah. info on May 15, "Insulting the Koran or the Prophet Mohammed is regarded as blasphemy and punishable by death in Saudi Arabia."

"Americans Interested in Koran After Abuse Scandal" headlined a June 21 article in the Saudi Gazette. The report said that the Council on American-Islamic Relations offered free copies of the Koran to Americans so that they can learn about Islam. The report added, "one copy was being requested almost every minute," and quoted the head of CAIR, Nihad Awad, as saying Americans want to better understand Islam.

As well, the lawyers for the 158 Saudi detainees at Guantanamo were quoted as saying that the Koran debacle could benefit their clients and could lead to their release.

The Saudi press was particularly critical. On May 20, Al-Watan cited historical "crimes" by America in Japan and Vietnam to suggest the Koran episode was only the latest in a long series of crimes - and that more should be expected. The article went on to suggest that America could make up for the Koran desecration if it were to "handover the criminal perpetrator to justice and let him stand in a cage similar to those built on Guantanamo."

With the debate surrounding the alleged desecration of the Koran at Guantanamo, it is easy to forget that those being held at the detention center are terrorists. If one were to believe the rhetoric coming from Saudi Arabia, its citizens held there are completely innocent and just happened to be in the wrong place at the wrong time. But as we learned on September 11, 2001, this is not the case.

N.Y. Imam Blames Zionists

July 6, 2005

Within weeks of September 11, 2001, an Egyptian sheikh serving as Al-Azhar University's representative in America and an imam at the Islamic Cultural Center and Mosque of New York City, Muhammad al-Gamei'a, was interviewed at an unofficial Al-Azhar University Web site, www.lailatalqadr.com. Sheikh al-Gamei'a returned to Egypt shortly after his interview was translated. Many of his friends in the interfaith community were shocked at what this "moderate" said.

In the interview, he claimed the Jews were behind the September 11 attacks and offered the following as evidence: "four thousand Jews did not come to work at the World Trade Center," and "Jews control decision-making in the airports and in the ... White House and the Pentagon." Another outrageous statement: "Muslims do not feel safe even going to the hospitals, because some Jewish doctors in one of the hospitals poisoned sick Muslim children, who then died." Sheikh al-Gamei'a also predicted the collapse of America.

Almost four years after Sheikh al-Gamei'a made headlines, another influential Egyptian-born New York imam gave a series of interviews to the press in his homeland. He also did not expect his words would be translated by MEMRI.

An imam at the Islamic Society of Mid-Manhattan and a lecturer on Islamic studies at Manhattanville College, Ahmad Dewidar, was recently called "the face of the next generation of Muslims in America." He has met with President Bush, Mayor Giuliani, Governor Pataki and U.N. Secretary-General Annan.

Recently, Dr. Dewidar attended the annual Conference of the Supreme Council of Islamic Affairs in Egypt, where he gave a series of interviews. In one that aired on MBC TV on June 9, he discussed the spread of Islam in America. He referred to sermons he had heard in 1995 that stated, "We are going to the White House so that Islam will be victorious, Allah willing, and the White House will become Muslim House." To view the broadcast, visit www.memritv.org.

On June 15, Dr. Dewidar was interviewed by the Egyptian Muslim Brotherhood's Web site, www.ikhwanonline.com. When asked about how the attacks of September 11, 2001, impacted the spread of Islam in America, he said that he witnessed hundreds of Americans converting to Islam. Hinting at an American government conspiracy related to the attacks, he said, "Whether or not these events were planned, or pinned on the Muslims, or something else - [it] provided an opportunity for [the American government] to legislate dubious laws that restrict the growth and presence of Islam in the U.S."

Regarding American skepticism toward Islam following September 11, 2001, he said, "The media - most of which is under Zionist control - has helped to spread this perception." Later in the interview he said that "the Jews" control the press. When asked, "What is the extent of the Muslim community's influence on American society?" He answered, "The Zionist community numbers only 3 million, but they control the government, the politics, the economy, and the media in the U.S."

Also while speaking to the Web site of the Muslim Brotherhood, he denounced President Bush's policy in the Middle East, claiming it was dictated by Natan Sharansky: "This Jew has despicable goals, and we see their effects today in America's actions in the region, imposing its opinion and its outlook on democracy, education, and political involvement on our countries."

Dr. Dewidar and Sheikh al-Gamei'a's interviews shared some themes, such as blaming Islam's poor image in America on a Zionist-controlled press. They also both said the Zionists control political decision-making and economic institutions. Both were extremely positive on the future of Islam in America and discussed a trend of Americans converting to Islam following September 11, 2001.

While the New York press has yet to pay this story adequate attention, it is being discussed by blogs such as Little Green Footballs. Others, such as Solomonia.com/blog, wrote about Dr. Dewidar's statements: "Before MEMRI - people could give these kinds of interviews abroad and no one would ever be the wiser. While still not broadcast widely, at least now, there's a chance of people wising up." On June 30, EconoPundit wrote: "MEMRI.org's most important contribution has been to make it harder for Middle East leaders to say one thing to their English-speaking audience while saying opposite and often contradictory things to their non-English-speaking constituents and allies. It now seems MEMRI's mandate has extended itself to U.S. Islamic leaders as well."

A Saudi's Millions in Gifts

July 13, 2005

A Saudi prince, Alwaleed bin Talal, who co-owns the ART TV network which includes the anti-American, anti-Semitic, pro-Jihad Arabic TV channel Iqra, has recently given millions to American groups, including $5 million to the Carter Center.

At an annual convention marking the 25th anniversary of the American-Arab Anti-Discrimination Committee (ADC) on June 1, which honored Rep. John Conyers, a Democrat of Michigan, and Rep. Nick Rahall, a Democrat of West Virginia, it was announced that the prince would donate $6 million to the ADC. Prince Alwaleed, who is the world's fourth richest person according to Forbes magazine, received the ADC's Global Achievement Award for his "outstanding humanitarian and philanthropic work."

Iqra first made headlines in 2002 when MEMRI translated a program featuring a 3-year-old Saudi girl named Basmallah calling Jews "apes and pigs" - she said it was a lesson from the Koran. The prince is currently involved in creating a new religious satellite channel that will make its debut later this year and can be expected to complement Iqra's content.

Available in America via satellite, Iqra bills itself as dedicated to bringing "the teachings of Islam into the homes and hearts of Arabs worldwide." According to its Web site, it offers religious, social, and economic programming, reaching viewers in the Middle East, Asia, Europe, and Africa. The station aims to explain concepts such as the ideological foundations of Islam.

Religious, political, and academic figures from throughout the Arab and Muslim world frequently appear on Iqra TV espousing hatred. Clips are available at www.memritv.org.

Over the past year, Iqra has broadcast lectures by leading Saudi clerics such as Aed al-Qarni, who encourages Muslims to gain victory over the West by "slitting of throats and shattering skulls." A former employee of the Saudi Embassy in Washington, D.C., Sheikh Al-Munajjid, said in an Iqra appearance that spreading Islam by the sword is justified, "the stupid acts of these Jews and Christians mobilize the

[Arab] nation," and "The big explosion will come!" Al-Azhar Sheikh al-Mussayar said on an Iqra talk show that Muslims should confront non-Muslims, and be aggressors - rather than defenders - of their religion against the Jews and Christians.

During another program, a reporter asked passersby on the streets of Saudi Arabia if they would shake hands with a Jew. All responded by cursing Jews. A soap opera called "Stories from before the Verses Came Down" portrayed Jews distorting the Torah and admitting that they live off the blood of prophets. The series also depicted the Jews refusing to acknowledge Mohammed's prophecy and called them "pigs."

Conspiracy theories surrounding the attacks of September 11, 2001, are also often broadcast on Iqra. An Egyptian historian stated on a talk show that the Vatican had delegated to America its mission to destroy Islam and that she also said that America itself carried out the attacks on assignment by the World Council of Churches. An Islamic preacher in California claimed in an interview that American authorities refused to investigate September 11 because America fabricated the event. And a former dean of Humanities at Cairo's Ein Shams University similarly asserted on a talk show that America was 100% behind the attacks.

Iqra programming frequently endorses suicide bombings in Israel. On one children's show, Sheikh Ahmad Qattan pleaded with Muslim parents to teach their children to become martyrs. "Oh mothers and fathers, you must train your children every night, before they go to bed, to go on raids in order to liberate Jerusalem and Al-Aqsa, and when he goes to sleep, after reading the Koran and the bedside verses, he should recite together with you the prayer for martyrdom," he said.

Prince Talal has also been praised for his commitment to women's rights. And Iqra has programming devoted to women's issues - much of it supports explicit violence against them. For example, on a weekly talk show, host Muhammed al-Mutwah has displayed different-sized rods to explain how to use them to beat viewers' wives. One was 10 feet long. It is unlikely that organizations such as former President Carter's Peace Center would have accepted funds from Prince Alwaleed if they were aware that he is also funding the kind of incitement against America broadcast on Iqra TV.

Must-See Saudi TV II

July 20, 2005

MEMRI hosted an event on Capitol Hill last June titled "Must-See Saudi TV." The event was sponsored by Senators Santorum, Schumer, and Collins and analyzed television programs broadcast by stations funded and controlled by the Saudi Arabian government, such as TV1, TV2, and satellite channels such as Al-Majd and Iqra TV.

Shows featured leading Saudi religious figures, professors, members of the royal family, government leaders, and intellectuals. Certain messages and themes surfaced again and again: calls for the annihilation of Christian and Jews; anti-American and anti-Semitic declarations; rallying cries for jihad and terrorism, and incitement against American troops in Iraq.

An hour before the event began, a Fox News reporter Sheppard Smith covering the meeting said on the air: "As terrorists threaten Westerners inside Saudi Arabia and everywhere else it seems like, a group of U.S. lawmakers plan to take a look at what makes the airways inside Saudi Arabia. They're gonna start watching TV there; in fact they're gonna get together to do that. The Middle East Media Research Institute will host some senators at a presentation of recently broadcast pieces from Saudi-controlled television. So what's the difference between what Saudis say in our TV boxes and inside their own?"

As I left the MEMRI office for the event, the phone rang. It was a frantic representative of the Saudi government press agency's office in Washington, D.C., asking what would be shown.

Tomorrow, a follow-up event will be held at the Capitol to examine what - if any - changes occurred at Saudi television outlets during the past year. The timing is notable since in February the kingdom hosted an international counterterrorism conference that produced the royal family's much-heralded "Riyadh Declaration." The communique emphasizes "the importance of enhancing the values of understanding, tolerance, [and] dialogue [and] becoming acquainted with other peoples, bringing cultures together, rejecting any clash of civilizations, and combating any ideology that calls for hatred, instigates violence, or justifies the terrorism crimes that are denounced by all religions."

Unfortunately, since September 11, 2001, Saudi Arabian officials have been inciting their own people and others in the Muslim world to hatred against non-Muslims, and in particular Jews and Christians. America has also been a target.

While compiling segments from Saudi TV to show on Capitol Hill tomorrow, countless clips had to be left on the cutting-room floor. It would have taken all day to view all the examples of hatred MEMRI gathered over the past year.

The hour-long compilation will include a talk show in which the audience laughs during a discussion of those murdered on September 11, 2001, and another wherein leading Saudi clerics explain why it is permissible to kill American soldiers in Iraq. The sampling also highlights programming on the coming collapse of America and a future Islamic conquest of the universe. Religious programs devoted to Koranic teachings encourage hatred of Jews and Christians, discuss the annihilation of Jews, and support suicide bombers by telling what happens when a martyr dies - they are rewarded by Allah with 72 black-eyed virgins.

Last week's Friday sermon from Islam's most holy site, the Al-Haram Mosque in Mecca, by its imam, Sheikh Abd Al-Rahman Al-Sudayyis, will be screened, too. In the sermon he stated: "Oh Allah, liberate our Al-Aqsa Mosque from the defilement of the occupying and brutal Zionists ... Oh Allah, punish the occupying Zionists and their supporters from among the corrupt infidels. Oh Allah, scatter and disperse them, and make an example of them for those who take head."

Previously, Sheikh al-Sudayyis's sermons have been characterized by confrontation against non-Muslims. He has called Jews "scum of the earth" and "monkeys and pigs" who should be "annihilated." Other enemies of Islam, he says, are "worshippers of the cross" and "idol-worshipping Hindus," who should be fought.

Any claims by the House of Saud that they are not spreading hate is simply a lie.

As the world witnesses Islamic-based terrorism, most recently in Britain, the global community must demand the Saudis stop their incitement. This is long overdue.

To view the clips shown on Capitol Hill tomorrow, a specially designated link can be found at www.memritv.org.

Arab Reformists Raise Voices on Terror

July 27, 2005

While the Arab world has seen a steady rise in the number of reformists speaking out against terrorism over the past four years, their statements have become increasingly urgent in the aftermath of the July 7 London attacks.

Within the Arab mainstream, two sides are battling for the future of Islam. One is the establishment, which includes regimes and their elitist supporters in the press, academia, mosques, and elsewhere. For years, they have used a mechanism that nurtures incitement against others to stay in power - without free elections - and encourage the Islamist movement now terrorizing the world. Their status quo positions are in jeopardy due to the outrage over the attacks in Britain.

On the other side is the reformist camp, which is fed up with the establishment. Its supporters are allying with the West and backing the struggle against ideological sources of terrorism. They include opposition political figures, student movements, intellectuals, authors, and columnists. Several statements by reformists who are active in the press follow.

In a July 13 editorial, the editor in chief of a Kuwati daily, Al-Siyassa, Ahmad Al-Jarallah, praised President Bush for his leadership in the war on terrorism and said that the London bombings gave him "more ammunition" to go after Al-Qaeda.

After criticizing those in the Arab world who defend terrorists, Mr. Al-Jarallah described the reformists' role, his own included: "Our interests go hand in hand with the modern world. We find ourselves on the same side as the civilized countries in the fight against terrorism." He described the reaction by victim countries as "a 'world war' against terrorism led by the West to defend its internal security. The war on terrorism is a war to protect human lives."

The editor in chief of the London Arabic daily Al-Hayat, Ghassan Sharbal, similarly called the events of July 7 a "chapter in World War III," which, he said, began on September 11, 2001.

In the article "Muslims Speak Out," the editor in chief of Asharq Al-Awsat, Tariq Alhomayed, encouraged Britain's Muslims to march in the streets against terrorism. On July 22 he wrote, "No to terrorism, not in our name! Instead of permitting the imams of the Internet and foreign satellite channels, to speak on your behalf. I urge you to follow the media in London because foreign channels, in your native tongues, encourage you to commit suicide, and extol the virtues of martyrdom."

Also in Asharq Al-Awsat, Mona Eltahawy wrote on July 17: "It is time to declare once and for all the absurdity of the 'George Bush made me do it' excuse that is dragged out every time Muslims carry out a terrorist attack. It was as ridiculous on July 7 when terrorists struck London as it was on September 11, 2001, when they hit New York and Washington. Many people across the world have opposed U.S. and British foreign policy, but they are not rushing to fly planes into buildings or to blow up buses and Underground trains in London ... we have to knock the phrase 'George Bush made me do it' out of our vocabulary. It is long past overdue that we stopped blaming everyone but ourselves. We have all known about the growing extremism and militancy among our communities, but it was easier to ignore them and say, 'We're not like that' than to confront it head on."

Al-Hayat's Washington correspondent, Salameh Nematt, wrote on July 18: "The terrorist targeting of Britain, after Spain, is not linked to the U.S. presence in Iraq. If the goal of the killers was to strike against the countries taking part in the occupation to force them to withdraw, then they would not have targeted Istanbul last year, since Turkey did not send troops to Iraq but opposed the war and forbade the U.S. forces from going into Iraq via its territories. The same applies to Saudi Arabia, Morocco, and Lebanon, which are all countries that opposed the war on Iraq but still are targeted by the same terrorism that seems to hit wherever possible."

Mr. Al-Jarallah, Mr. Alhomayed, Ms. Eltahawy, and Mr. Nematt are just a few of the many in the reformist camp. To read more about other reformists and their cause, visit www.memri.org/reform.html. Their success is intertwined with our safety and, ultimately, whether we win or lose the war on terror.

CAIR Silences Its Critics

August 3, 2005

Since September 11, 2001, the Council on American-Islamic Relations has used subsequent terrorist attacks to educate Americans about Islam through radio and TV.

Following the London bombings, a Washington, D.C.-based radio host, Michael Graham of the Disney owned 630 WMAL, devoted a show to the attacks, saying that followers of Islam were terrorizing Britain. Some of Mr. Graham's statements - such as "Islam is at war with America" and "Islam is a terrorist organization" - resulted in a campaign by CAIR to get him fired.

On July 25, CAIR's head for southern California, Hussam Ayloush, appeared on Hugh Hewitt's syndicated radio show. As a blog run by former Navy SEALs, Froggy Ruminations, reported, "For the first two hours of his program today, Hugh unwittingly turned the keys of his radio show over to an organization that has a long track record of apologizing for Islamic terror." The blog said that Mr. Ayloush specifically criticized MEMRI. Its writers added, "Ayloush stonewalled on every point, and gave the impression, like a good little CAIR spokesman, that there really isn't any terrorism in the world."

"I can't remember all the deceptions and outright lies that were told and many of them went unchallenged" is how Froggy Ruminations summed up the interview. Those words could also apply to a July 25 interview between a CAIR spokesman, Ibrahim Hooper, and a sympathetic Chris Core, who hosts a radio show on the same station as Mr. Graham. Mr. Hooper was brought on the air to discuss statements Mr. Graham made and used the opportunity to further CAIR's ongoing unofficial Da'wa campaign of Islamic missionary activity to teach Americans about Islam and to influence them to convert. In some ways, CAIR has already been successful. They openly state in Arabic that tens of thousands of Americans have converted to Islam, an unsubstantiated claim.

During the interview, Mr. Core asked Mr. Hooper to address Western criticism of calls for Muslims to kill infidels found in the Koran. Mr. Hooper said "that was nonsense," explaining that while there are such verses, they only refer to military campaigns during the time of Mohammed. This does not jibe with the views and teachings of leading Islamist scholars throughout the world.

As the Saudi-based Web site Al-Minbar quoted Sheikh Muhammad Saleh al-Munajjid as saying: "Muslims must ... educate their children to Jihad. This is the greatest benefit of the situation: educating the children to Jihad and to hatred of the Jews, the Christians, and the infidels; educating the children to Jihad and to revival of the embers of Jihad in their souls."

The program got interesting when Mr. Core asked Mr. Hooper to explain the concept of Jihad. He said, "Jihad is a respected concept in Islam. Jihad does not mean flying planes into buildings. It does not mean blowing up subways ... I just read a Tradition of the Prophet, it's called a hadith ... It said, 'the best Jihad is to speak a word of truth to an oppressive ruler' ... Jihad in its military sense means a Muslim nation having a standing army, people in uniforms there to defend the borders, to prevent invasions, to do anything any normal army does, for any nation. That's what a real Jihad is. It has to be called by a legitimate government... It's like having the U.S. army only it's in the Muslim context."

Mr. Hooper complained that jihad is only reported on when it is referred to as "holy war," which has been "misused by people on both sides of the spectrum." The way in which Mr. Hooper and CAIR define jihad to Americans is misleading. According to the Islamic Affairs Department of Saudi Arabia's embassy in Washington, D.C., one definition of jihad is the following: "The Muslims are required to raise the banner of Jihad in order to make the Word of Allah supreme in this world, to remove all forms of injustice and oppression, and to defend the Muslims. If Muslims do not take up the sword, the evil tyrants of this earth will be able to continue oppressing the weak and [the] helpless."

CAIR & Co.'s Fatwa

August 10, 2005

A much-heralded Fatwa or Islamic religious ruling "against terrorism" was issued on July 28 by Muslim academics in America and Canada. The edict was signed by 18 scholars from the Fiqh Council of North America and was endorsed by more than 120 Muslim groups. The Islamist organization Council on American-Islamic Relations was given the honor of officially releasing the statement.

"Bogus" is how Steve Emerson of the Investigative Project described the Fatwa on "The Counterterrorism Blog": "Nowhere does it condemn the Islamic extremism ideology that has spawned Islamic terrorism. It does not renounce nor even acknowledge the existence of an Islamic Jihadist culture that has permeated mosques and young Muslims around the world. It does not renounce Jihad ... In short, it is a fake fatwa designed merely to deceive the American public into believing that these groups are moderate. In fact, officials of both organizations have been directly linked to and associated with Islamic terrorist groups."

The Fatwa was heralded by major news outlets, which have grown accustomed to using CAIR's press releases without any scrutiny. In an interview with the Saudi weekly Ukaz in 1999, CAIR leaders discussed the organization's manipulation of the press under the headline "A Muslim America - Only a Matter of Time." One leader stated openly, "We control the media instead of the media controlling us." CAIR's founder, Nihad Awadh, described techniques utilized by the group's spokesman, Ibrahim Hooper, to infiltrate the press: "You will be surprised to hear that all his articles do not go through censorship and correction in the big papers," Mr. Awadh said.

A CAIR leader added, "CAIR [exerts] today an organized and educated effort, using American media techniques and exploiting American legal loopholes ... For example: One of the editors of the American monthlies attacked the Muslims ... Answering these allegations, CAIR convened a press conference and recruited all Muslims to call the monthly in every communication method, and we paralyzed the paper's interests, and the man [the editor] was forced to apologize ... We exploit this position whenever someone else rises up to attack Muslims, and present to him what happened to those before him, and then he retracts ... This method has gained us great achievements and stops serious attempts to attack Islam."

A New York Times story about the Fatwa quoted CAIR, as well as Jamal Badawi, who was identified as chairman of the Islamic Foundation in Halifax, Nova Scotia. Mr. Badawi is influential not only in Canada but also in America. He has been a guest lecturer at Michigan's "The American Learning Institute for Muslims"; directs the Islamic Information Foundation, which seeks to promote better understanding of Islam by both Muslims and non-Muslims and is a federally registered charity in Canada and America; serves as a faculty member at St. Mary University in Maryland, and used to be a lecturer at Stanford University in California.

In an interview with the Saudi Gazette on June 24, 2005, while visiting the kingdom, Mr. Badawi said: "9/11 was un-Islamic," and "I strongly condemn the September 11 attacks ... whoever did it," adding, "It is not confirmed yet who is actually behind the attacks." In the Saudi Gazette story, Mr. Badawi discussed the difficulty in drawing up an "agreement on how terrorism should be defined" and said exploding cars in Iraq's markets and killing civilians should be included.

Mr. Badawi went on to make an outrageous claim - that America could be behind these attacks: "This has to be investigated as to who is actually behind this ... There have been allegations that I cannot confirm that people going to the market to buy vegetables are stopped in the name of inspecting their cars by [American] forces, their hands are tied and they are blindfolded. There have been cases and I want a clarification from American officials to these allegations. After inspecting their cars they are allowed to go and when the car reaches [the] checkpoint it explodes and they call them suicide bombers, perhaps the occupants of the car were not even aware that they are carrying a bomb in their car. Such incidents should be thoroughly probed."

Mr. Badawi was number seven of 18 to sign the Fatwa. Other signers have reported connections to terrorist organizations. If this is the best response leading American Muslim leaders can do to fight terror, then we have a problem in America.

The Morning After Gaza

August 17, 2005

The Road Map for peace called for an end to Palestinian armed activity, including for the Palestinian Authority to collect arms from terror groups and for all official Palestinian institutions to end incitement against Israel. None of which has happened. Now that Israel's disengagement from Gaza is a fact — a major question remains, what will happen the day after?

All indications point to Gaza becoming a haven for gathering Palestinian terrorist groups, as well as global terrorists affiliated with them. Many have openly stated they will be moving their bases of operations there.

The Saudi daily Al-Watan reported in the last week of June that Palestinian Authority leaders have invited all Palestinian rejectionist groups to Gaza. They include multiple groups and individuals on the U.S. list of Designated Foreign Terrorist Organizations including Syrian based Hamas officials such as Khaled Mashaal and Musa Abu Marzouk, as well as leaders from the Democratic Front of the Liberation of Palestine and the Popular Struggle Front.

It was also announced that the Tunis based head of the PLO, Faruq Qaddumi is moving his base of operations to Gaza. He is a possible successor and key rival to PA Chairman Mahmoud Abbas. According to him, the PLO charter which calls for the destruction of Israel has not really been amended. Qaddumi will be bringing with him a 'volunteer popular army' of at least 1,500 who will be under his control according to Saudi media reports from August 3.

This follows other reports from Lebanon that additional Palestinian fighters devoted to Israel's destruction are also coming to Gaza. One report, in the Lebanese Daily Star from August 1st quoted a senior Fatah official who "did not deny rumors" and was quoted explaining "we were informed we will be with them [other Palestinian fighters in Gaza] soon."

Additionally in May, Arabic papers such as Al-Ayyam and Al-Quds Al-Arabi reported that hundreds of Palestinians wanted for connections to terrorist attacks have joined the Palestinian security apparatus. Other jailed terrorists have been released from prison this summer. According to the Arab News of June 12, members of Islamic Jihad who were involved with the suicide attack in February that killed five in Tel-Aviv have been sent free.

The issue of arms remains contentious as the U.S. is urging Israel to re-arm the PA while terrorists who were supposed to be disarmed under the Road Map have not yet been disarmed. In June, the PA's Foreign Minister Nasser Al-Kidwa, Yasser Arafat's nephew, explained on PA TV: "Keeping our weapons is a strategic option" and that "possession of weapons is a strategic issue." Referring to terror groups such as Hamas and Islamic Jihad, he explained "the dismantling of armed organizations is not on the table because weapons are legal."

In the days and hours leading up to disengagement — the PA leadership has been stating in Arabic how it views the situation. On August 12, Abbas exclaimed "Today were are celebrating the liberation of Gaza and the Northern West Bank, tomorrow we celebrate the liberation of Jerusalem."

Similarly, during the first week of August, PA Prime Minister Ahmad Qurei in celebration of Yasser Arafat's birthday exclaimed, "Today Gaza, tomorrow Jerusalem" adding: "We will still celebrate once we reach our goals and establish a Palestinian state with Jerusalem as its capital, with the return of the refugees."

Recognizing the role of Hamas, Qurei stated: "the children of Yasser Arafat the martyr and the children of Ahmad Yassin the martyr and children of all the martyrs. Jerusalem is closer than ever... We will continue until victory. None of us will give up until we are on Jerusalem's soil."

The leadership of Hamas has been the most open about its intentions. According to Al Jazeera.net on August 12, the groups leader, Mahmoud Al-Zahar, explained his organization is a legitimate security tool and will never disarm. After attending a training session of the military wing of Hamas, the Izz Al-din al-Qassam Brigades, he stated the Palestinians will "liberate Palestine from the Jordan River to the sea."

The armed wing of the Fatah movement, which is a part of the PLO issued a statement on August 15 that sums up the problems ahead: "We support our brothers in Hamas and affirm that no Palestinian armed wing will be dissolved because Palestine is not Gaza only, but rather, Palestine is Palestine from the River (Jordan River) to the Sea (Mediterranean Sea) and it is our land and we have the right to defend it."

Arafat's Heir Apparent Comes to Gaza

August 24, 2005

As this column reported last week, the Tunis-based head of the Palestine Liberation Organization, Farouk Qaddumi, whose official position on Israel is that it should be destroyed, will move into Gaza this summer. He is a possible successor and key rival to the Palestinian Authority's current leader, Mahmoud Abbas. Mr. Qaddumi will be bringing with him a "volunteer popular army" of at least 1,500, according to Saudi press reports from August 3.

Mr. Qaddumi confers often with terrorist groups and regimes. At a September 4, 2003, meeting in Lebanon with the head of Hezbollah, Shiekh Hassan Nasrallah, Mr. Qaddumi announced "a cohesion between the Lebanese and Palestinian resistance." Over the last decade, he repeatedly met with Hamas and Islamic Jihad leadership in Damascus to plan "the liberation of Palestine."

He also has served as a point person between Palestinian Arabs and the Iranian regime. In an interview from Iran with Al-Jazeera broadcast on December 20, 2004, Mr. Qaddumi noted that Yasser Arafat was the first Arab leader to visit Iran after the Islamic Revolution in 1979 - a slight directed toward America. The PLO chief said on-air that the purpose of his visit to the Islamic Republic was to strategize with the regime on the "liberation of Palestine."

When asked if it was true "that Iran is infiltrating the Palestinian ranks, as far as Gaza, through the Hezbollah Cadres, or the Palestinian Islamic organizations in Iran," he responded that it was "good" and "truly positive" and that the PLO "welcomed all Arab and Islamic organizations to infiltrate." In a December 12, 2004, interview on Iran's Al-'Alam TV, he said the "cooperation between the PLO and Iran" includes working "to liberate Jerusalem."

Mr. Qaddumi was also the Palestinian-Arab link between the U.N. oil-for-food program and Saddam Hussein's regime. The Iraqi News Agency reported on August 16, 2002, that Mr. Qaddumi visited Iraq and met with the chairman of the Iraqi Revolutionary Leadership, 'Izat Ibrahim, who was immortalized as the eight of diamonds on the U.S. military's Terror Card Series of Wanted Iraqis. While there, Mr. Qaddumi said: "The battle of Iraq and [the battle of] Palestine are one battle against their historic enemies. The martyrs in Iraq and in Palestine are martyrs of the first rank, as they fight, with their pure, noble souls, against the ones who conquered their land, their honor, for their survival, their history, and their civilization ... against the most arrogant force of evil, oppression, and aggression on the face of the earth."

In talks between Mr. Qaddumi and Saddam in Iraq on January 18, 2001, the Iraqi dictator promised "Iraq is ready to bombard Israel for six continuous months of artillery from land and sea."

Mr. Qaddumi has made multiple, on-the-record threats to America, calling for terrorist acts against American interests. He called the attacks of September 11, 2001, "a lesson to the U.S." saying it "was the first time that Arabic names entered every American household." Mr. Qaddumi has visited America, but it was reported on August 23, 2003, that he was denied permission to speak at the United Nations.

The PLO charter calling for the destruction of Israel was reportedly amended in 1996, but on an appearance on ANB TV on April 25, 2005, Mr. Qaddumi said that he "does not accept any amendment." Furthermore, in an interview with an Israeli-Arab weekly on January 3, 2003, he openly stated that there is no difference between Hamas and the PLO vis-a-vis calling for the destruction of Israel. He said, "strategically we are no different from it."

One of Mr. Qaddumi's most popular phrases this past year has been, "There are 300 million Arabs, while Israel only has the sea behind it." He has added, "Time is on our side." As he said at a conference of foreign ministers of Islamic states held in Qatar on December 12, 2001, "The countdown [to the end] of Israel will begin ... and will find its end as the USSR and Yugoslavia did."

Top 10 Arab and Iranian Conspiracies

August 31, 2005

This summer, foreign ministers from 57 Islamic countries met in Yemen to devise a strategy for a summit of the Organization of the Islamic Conference to be held in November at Mecca. A communique warned "the international community to the dangers deriving from Zionism, neo-conservative ideology, aggressive Christian Evangelism, Jewish radicalism, Indian radicalism, and secular radicalism."

Details of such "dangers" frequently appear in the Arab and Iranian press and have been documented in past columns. The following highlights their Top 10 conspiracy theories of the summer:

10. On June 30 a London-based, Arabic daily, Al-Quds Al-Arabi, quoted a conservative member of Iran's parliament, Shakir Allah Attar Zadeh, as saying that Secretary of State Rice's hostility toward his country stems from her Iranian boyfriend leaving her during her schooldays. Similarly, an editorial in the Tehran Times on July 14 stated that America is cooperating with the Taliban and is behind the increase in drug smuggling in Afghanistan.

9. Palestinian Arab leaders - such as Farouk Qaddumi in the Tehran Times of July 13 - have continued to accuse Israel of poisoning Yasser Arafat. The commander of Arafat's Presidential Guard, Munir Al-Zu'bi, reiterated the claim on August 13 on Al-Arabiya TV.

8. In another London Arabic daily, Al-Hayat, writer Adel Malek on June 28 allegedly revealed the contents of a secret "U.S.-Israeli memorandum" that detailed a plan to strike Syria and Lebanon, "eliminate Assad," and arrest Hezbollah's leader, Nasrallah. Similarly, an Arab novelist and political analyst, Sherif Hetata, wrote in the Egyptian weekly Al-Ahram of June 23-29 about a planned "joint U.S.-Israeli strike against Iran in the works." Mr. Hetata called for a grassroots movement to unseat "war criminals in the U.S., Britain, and Israel."

7. Following the assassination of Egypt's ambassador to Iraq, a former Egyptian ambassador to Afghanistan, Ahmad El-Ghamrawi, in an interview with Islam Online blamed the Mossad for the crime. A member of Lebanese parliament, Marwan Fares, was quoted on June 24 in the Syrian government daily Teshreen as blaming the Mossad for the assassination of the secretary-general of the Lebanese Communist Party, George Hawi.

6. An July 11 article in the Tehran Times said that America had not captured Osama bin Laden or Ayman al-Zawahri to keep Americans scared of terror attacks and to allow the war to continue. Similarly, an article in the Turkish daily Zaman on July 9 stated: "Al-Qaeda rather serves Western imperialist goals ... If the Al-Qaeda factor had not emerged, would American expansionism have found so much legitimacy, at least within its own public opinion."

5. In Egypt, a debate has surrounded the translation of a 19th-century biography of Mohammed by the author "Reverend George Bush." The book called Mohammed "an imposter" and as the Al-Ahram Weekly reported in July, called Muslims "locusts." According to the State Department, the author of the book is not a direct ancestor of the president, as the Egyptian press asserted.

4. Following reports earlier this year that the Sudan genocide was a Jewish conspiracy, an article on July 22-23 in the London Arabic daily Azzaman made similar, though more detailed, claims.

3. A book reviewed by Hassan Hanizadeh in the Iranian daily Mehr News of July 3, on the history of the Jews included a chapter on how "certain Jewish rabbis distort the Old Testament." Similarly, an article by Hasssan Tahsin in the Saudi Arab News of July 8 discussed Russians in Israel: "Russian Jews can't be buried in Jewish cemeteries ... Jewish families are not allowed to host Russian Jews ... A Russian may be killed if he was found speaking Russian ... Parents teach their children to refer to Russian Jews and Christians as pigs and gentiles."

2. The bombings at Sharm Al-Sheikh on July 23 were blamed on Israel - including by Mustafa Bakri in Al-Asbu', who pointed to a "Mossad branch in Baghdad," and an attorney for Egypt's Supreme Court of Appeals, General Osama Halawa. On July 25, the Iranian daily Kayhan International blamed the "Mossad operating behind the camouflage of Al-Qaeda" and wrote that "the Zionists" were also behind London's bombings.

1. The London bombings were the subject of the most wide-ranging conspiracies this summer. The Mufti of Mt. Lebanon, Dr. Muhammad Al-Jozo. blamed the Zionists on ANB TV on July 24. Iran's minister of information, Ali Younesi, was quoted by IRNA as pointing to "U.S. and Israeli Intelligence" who have "infiltrated Saudi Arabia, Palestinian, and even European States" including London.

World War III: Four Years and Counting

September 7, 2005

The war against the West - which was brought to America on September 11, 2001, and has continued throughout the world with attacks in Spain and London - has been referred to by leading Arab writers as "a chapter in World War III." Terrorists have beheaded and slit the throats of Westerners and blown up planes, trains, and buses in purposefully brutal acts intended to spread fear. One notable example includes the suicide car bombing of a crowd of children receiving candy from America soldiers in Iraq, killing 24 of them on July 14, 2005.

As the fourth anniversary of September 11 approaches, it is worth reviewing what President Bush told the American people on September 15, 2001: "This is a conflict without battlefields ... a conflict with opponents who believe they are invisible ... Victory against terrorism will not take place in a single battle, but in a series of decisive actions ... You will be asked for your patience, for the conflict will not be short ... You will be asked for your strength, because the course to victory may be long."

It is also important to remember that the attacks on the Pentagon and the World Trade Center took a long time to plan and implement. Now Al-Qaeda members have vowed that more attacks are on the way. A tape from Osama bin Laden's deputy, Ayman al-Zawahiri, was broadcast on Al-Jazeera on August 4, "Oh, Americans, what you have seen in New York and Washington ... are nothing but the casualties of the initial clashes."

Other Al-Qaeda leaders have warned of future attacks. Islamist Web sites in June 2002 quoted the group's spokesman at the time, Suleiman Abu Gheith, as saying that Al-Qaeda will be "continuing our blows against the U.S." and that "America must prepare itself ... Allah willing, the blow will come from where they least expect it." He warned: "We have the right to kill 4 million Americans - 2 million of them children - and to exile twice as many and wound and cripple hundreds of thousands."

Several Islamist Web sites in November 2003 posted an interview with a close associate of bin Laden, Abu Salma al-Hijazi, in which he promised a "huge and very courageous strike" by Al-Qaeda. He predicted the number of "infidels" killed would exceed 100,000, warning: "We are patient," and "our patience will only end with the collapse of America and its agents."

Mahfouz Walad al-Walid, otherwise known as Abu Hafs "the Mauritanian," was interviewed on November 30, 2001, on Al-Jazeera. "One of the acts of grace of this generation was to kill the Americans, to incite to this killing, to fight in the jihad against them with full force. I and my brothers in the Al-Qaeda organization ... swore an oath to carry out the mission," he said, adding, "In our opinion, America has entered the phase of the beginning of the end."

Al-Qaeda's stance on the destruction of America is shared by many prominent Islamic scholars who are frequently quoted by the Arab press. A Saudi professor, Nasser Bin Suleiman al-Omar, spoke on Saudi-owned Al-Majd TV on June 13, 2004: "Today, America is defeated. I have no doubt, not even for a minute that America is on its way to destruction. But as Ibn Khaldoun said, just as it takes decades for nations to rise, it takes them decades to collapse. They don't collapse overnight ... It will be destroyed gradually. America will be destroyed. But we must be patient."

The volume of calls for America's collapse and the ongoing celebration of September 11, 2001, are rampant. Sheikh Mahmoud Al-Masri who appeared on Al-Majd TV on March 2, 2005, explained what it was like being in New York on that day to an Arab audience who laughed with him when he described "how beautiful" it was as the World Trade Center fell and the reaction of the "wretched" American people who quickly embraced Islam.

As the war on terror continues, action must be taken by Western governments. Supporting true Arab and Muslim reformists both here and abroad is crucial. To date, MEMRI's work has introduced many reformists to the West, as well as led to shutting down Islamists Web sites hosted in North America; exposed extremist clerics who say one thing in English and another in Arabic, and led the way in banning Arab and Iranian TV channels such as Iran's Sahar TV and Hezbollah's Al-Manar throughout the West. There is much unfinished work to complete as the war on terror continues.

Musharraf, Mohammed, and the Jews

September 14, 2005

The annual United Nations General Assembly serves as an opportunity for world leaders and diverse groups to meet. One such meeting this year includes Pakistan's president, Pervez Musharraf, reaching out to American Jewish groups.

On September 19, 2001, Mr. Musharraf made a televised message to the Pakistan people to explain how his nation would respond to September 11. Referring to "the first six years of the history of Islam," Mr. Musharraf said: "Mohammed himself went from Mecca to Medina to save Islam ... when the prophet reached Medina, then he entered a friendship treaty with his enemies, the Jews...because of the peace treaty with Jews, he was able to fight the infidels of Mecca. After six years, the Jews noticed that Islam was becoming stronger. They got very irritated."

Mr. Musharraf then explained why Mohammed was willing to sign a peace agreement with the infidels even if its temporarily made Muslims look weak." There was a battle between Muslims and Jews, and of course by the grace of God, Muslims won the battle ... Mecca was conquered. What lesson do we derive with this incident? ... On this occasion, a strategic decision is to be taken...if you are facing two problems and you have to choose one, then it is better to take the lesser evil...."

The temporary good relations between Mohammed and the Jewish tribes of Arabia (Qurayza, Khaybar, Khaynuqa, and Nadhir) who were wiped out by Muslims is well-known in the contemporary Muslim world, as it is often mentioned in text-books, sermons, and TV shows. Yasser Arafat frequently referred to the concept of a temporary truce in justifying the Oslo Accords with Israel. Other Arab and Muslim leaders also frequently refer to this.

For example, following threats to bomb Los Angeles, Egyptian labor party head Magdi Ahmad Hussein appeared on Al-Jazeera on October 17, 2004, explaining that "600 to 700 prisoners were killed in the raid on the Qurayza tribe. Why do they conceal this? Why do they hide the fact that the Prophet gave the order to assassinate some poets - to assassinate! Not in military operations, but rather by individual assassination. Why did he order the assassination of K'ab Ibn Ashraf, the Jew, leader of Khaybar ... As a result, the Jews became fearful and terrified."

Syrian legislator Mohammed Habash appeared on Syrian TV on July 29, 2005, discussing the characteristics of the Jewish tribes of Arabia. He explained that the Jews smelled so bad it "made people sick" and that "when the Prophet's companions reached Al-Madina, they all contracted fever because of the filthy environment."

"Stories From Before the Verses Came Down" was a Jordanian-produced series that ran on Saudi Iqra TV in February 2005; which was based on the Jews of the time of Mohammed. The series portrayed the Jews as despicable "pigs" scheming against Mohammed. Part of the plot included Jews altering the Torah to hide the fact that Mohammed was a true prophet; how early Jews planned "a very long war" against the decedents of Muslims and Jews depicted as having "joined forces with Satan." One episode showed Jews engaged in voodoo, creating a doll of Mohammed and sticking it with needles in his head, liver, and eyes.

Sheikh Yousuf Al-Qaradhawi, one of Sunni Islam's most prominent scholars, explained on his Al-Jazeera show on January 19 this year: "One of the first things the Prophet Mohammed did when he went to Al-Madina was to make a pact with the Jews ... The problem started because of the Jews. They were the ones who were treacherous and violated the pact." Mr. Al-Qaradhawi added, "The problem with the Jews is not one of faith or religious laws. The problem is the covetous aspirations that have characterized their attitude since the days of the Prophet Mohammed, that brought this position toward them since the days of Mohammed, and later on in our times."

By reaching his hands out in friendship to the Jews, President Musharraf has a unique opportunity to distinguish himself from other Arab and Muslim leaders who have given messages of peace in English and hate in Arabic.

Saudi Arabia and the WTO

September 21, 2005

The office of the American trade representative announced on September 9 that America and Saudi Arabia have concluded a bilateral agreement for the kingdom to join the World Trade Organization. Once the technicalities are resolved, the kingdom will become an official member - something they have been working for since 1993.

During this time, there has been opposition by members of Congress. For example, a May 25 letter signed by 47 House members to the American trade representative, Robert Portman, outlined areas of concern regarding Saudi policies, including the lack of full cooperation in combating terrorism.

The letter quoted officials from the Treasury Department criticizing the kingdom for "not establishing an oversight commission to regulate its charitable sector" as well as "for its failure to set up a Financial Intelligence Unit." The letter ended by urging the "administration to oppose Saudi accession to the WTO" until it takes more "meaningful measures to cut off both the finances and ideologies that feed terrorist movements across the globe."

This has not occurred. For example, a program on Saudi Iqra TV on August 29 was devoted to supporting jihad in Palestine. The program began with the host's telling all Saudis including "women, children, elderly, and youth" that they should donate money and how to do so.

A caption then appeared on the screen: "Saudi Committee for Support of the Al-Quds Intifada, Account #98, a joint account in all Saudi Banks." A moderator explained, "Jihad is the pinnacle of Islam" and that money will go directly to those waging jihad and "helps them carry out this mission."

In 2001, Account #98 was cited in an official Saudi government report that extensively detailed Saudi financial support to Palestinian Arabs, including families of suicide bombers. According to the report, the Saudi government pays between 10,000 and 20,000 Saudi royals to the families of martyrs, noting at the time of the report that 986 families had received funds from the account that were passed from Saudi Arabia's Islamic Development Bank directly to the Unified Treasury of the National Palestinian Authority account.

For decades, the Saudi royal family has been the main financial supporter of Palestinian Arab terrorist organizations. According to records on a Saudi royal family Web site, Ain-Al-Yaqeen, the Popular Committee for Assisting Palestinian Mujahideen, and the Support Committee for the Al-Quds Intifada and the Al-Aqsa Fund have given more than $4 billion and reportedly pledged Palestinians up to $1 billion to finance the continuation of the Intifada, which they commonly refer to as "jihad" and "resistance." The committees are headed by powerful Prince Salman Ibn Abd Al-Aziz, the governor of Riyadh, and Prince Nayef bin Abd Al-Aziz, the interior minister.

The August 29 Iqra program included the secretary-general of the Saudi government's Muslim World League Qur'an Memorization Commission Sheikh Abdallah Basfar, explaining why it was an "obligation" for all Muslims to support jihad.

Sheikh Basfar promised that "all of the funds sent via the known charities and organizations" will reach those for whom the funds are meant. The Saudi government official then cited a few Hadiths, or sayings of Mohammed, including: "He who equips a fighter - it is as if he fought himself." Another Hadith explained: "Someone who does not fight or equip a fighter - Allah will afflict a disaster upon him."

Sheikh Basfar's statement explaining "money is so important ... jihad can not be waged without it," was followed by the question, "Who will care for the families of these martyrs who sacrifice their souls? Can we be stingy with our money?"

The Saudi committees are cited in the congressional Saudi Accountability Act. This legislation calls on the Saudis to "permanently close all charities ... in the kingdom that fund, train, incite, encourage, or in any other way aid and abet terrorism anywhere in the world. "It also calls for an "end of providing support for the families of individuals who have committed acts of terrorism."

Should the Saudis not comply with the requirements named in the Saudi Accountability Act, the proposed legislation stipulates that sanctions are to be placed against them.

HBO Weighs In on 'Protocols'

September 28, 2005

This month marks the 100th anniversary of the first printing of the infamous anti-Semitic forgery, "Protocols of the Elders of Zion." The anniversary will be commemorated throughout the world including a two-day conference at Boston University's Elie Wiesel Center for Judaic Studies. Tonight in New York, HBO Films will premiere the documentary "Protocols of Zion" by Marc Levin's Blowback Productions (www.blowbackproductions.com). The documentary will be screened later in the month in selected theaters throughout America.

"The Protocols" are frequently referred to in the Arab and Iranian press connecting Jewish conspiracies to, among other things: assassinations of Arab leaders; the war in Iraq; control of Hollywood, the press, economics, and world resources; and an alleged war on Islam. As Mr. Levin's film illustrates, the most popular conspiracies are related to Jewish involvement in the attacks of September 11, 2001.

For example, the president of the Middle East Center for Studies and Public Relations, Hisham Jaber, appeared on Hezbollah's Al-Manar TV on July 11, 2005, stating: "I believe the events of 9-11 were not ... perpetrated by Al-Qaeda alone ... it is global Zionism that stands to gain the most from this ... we know that since the Protocols of the Elders of Zion, Zionism has forged the New Testament - and by now, 60 million in the U.S. alone have left Christianity to become believers in the Torah. Global Zionism has tried to forge the Holy Koran."

The secretary-general of the Islamic Universities Association, Jafar Abd Al-Salim, also appeared on Hezbollah TV on April 22, 2004: "The Protocols of the Learned Elders of Zion clearly refer to this: 'Spread corruption amongst all human societies so you can rule.' ...All the pornographic films in the world are made by Jewish companies, and the same goes for drugs."

Sheikh Muhammad Al-Mussayer of Egypt's Al-Azhar University, the most prominent school of Sunni Islam, spoke on Saudi Iqra TV on July 12, 2004: "What do they have to say about the perverse ideology of 'The Protocols of the Elders of Zion?'... What do we have to say about Zionism penetrating Christianity these days?

Christianity and Judaism have a long history of animosity...This is not love but a relationship whose goal is to destroy Islam."

Leading Palestinian Authority religious figures often cite the "Protocols." Saudi Al-Majd TV aired an interview on February 20 with Palestinian Arab Mufti Ikrima Sabri on the assassination of Lebanese leader Rafiq Al-Hariri: "'The Protocols of the Elders of Zion' and specifically the Talmud will discover that one of the goals of these protocols is to cause confusion in the world and to undermine security throughout the world." Another example includes a Palestinian Authority TV Friday sermon by Sheikh Ibrahim Mudeiris on April 16, 2004: "Our battle with our enemy is great ... We face a battle against the Crusaders and against World Zionism, which was exposed by the French intellectual, Garoudy, some 20 years ago ... He converted to Islam and wrote of World Zionism's covetous aspirations, based on 'The Protocols of the Elders of Zion.'"

A Muslim cleric from Denmark, Nabil Shaker Al-Taleqani, gave an interview on October 30, 2004, to Iran's Al-Alam TV explaining: "Go back to the book 'The Protocols of the Elders of Zion.' We must return to the 24 protocols, to these secret documents. When we turn the pages of these protocols, we see there are some people who pretend to be Muslims, but are actually Jews, and they are the ones who began to distort Islam. These people are behind this media, which is entirely against Islam ..."

Last year the Islamic Republic of Iran Broadcasting's channel directed at America, Jaam-E-Jam 2, broadcasted a series dealing with the Jews and the film industry, which centered on the "Protocols" detailing "the satanic Jewish ideas of taking over the world."

The Iranian news channel IRINN TV also aired a program about a book published on the "Protocols" by the Organization for Culture and Arts of the Tehran municipality on July 14, 2004: "These 'Protocols' [the Jews] emphasized the role of economics and the need to control it in order to advance their goals ... They believed that in order to rule the world they must have two tools in their control. The first is gold ... and the second is the international media ... In another place [in the 'Protocols'], they note the following solution to deal with countries that resist Zionism."

Mr. Levin's documentary will make an important contribution to all future historical studies of the "Protocols."

Ramadan as the Month of Jihad: A European Problem

October 5, 2005

In the modern era, the month of Ramadan, especially for Muslim communities in the West, is a month of religious spirituality. In Muslim tradition, however, it is also associated with being the month of jihad, the month in which Allah grants military victories to his believers.

This concept is often mentioned in Friday sermons and education curricula of the Middle East. For example, the latest Palestinian Authority schoolbooks written by Palestinian education experts remind the children that during Ramadan Muslims triumphed in battles throughout Islam's history. An Islamic education textbook for sixth graders includes a lesson about designating Ramadan as the month of "great victories in Islamic history," a "month of strength," and a month of jihad. The lesson explains that the ultimate proof of this month's importance is the "renowned Ramadan War of 1973 between the Arabs and Israel." The emphasis is not on the fasting, but rather on the jihad. Other battles are also mentioned: the battle of Badr in 624, the conquests of Mecca in 630 and Andalusia in 711, the battle of Al-Zallaqa (in Andalusia) in 1086, and the battle of Ein Jalut in 1260, as well as the 1973 War.

In an article titled "Muslim Victories During Ramadan" Fuad Mukheimar, an Al-Azhar University lecturer and secretary-general of the Egyptian Sharia Association, wrote about this during the Ramadan following the attacks of September 11,2001: "The significant battles during the month of Ramadan in which Allah granted victory to Islam and Muslims were...the first battle between believers and infidels. The Muslims demonstrated rare heroism, setting their souls on the points of their lances and preferring martyrdom to remaining alive."

From "The Mother of All Battles" mosque in Baghdad on Iraqi TV during Ramadan 2002, Sheikh Bakr Abd Al-Razzaq Al-Samaraai gave a Friday sermon one month before the war: "...We need the grace of jihad of the soul ... particularly in this difficult hour in which the Islamic nation [is] experiencing ... the challenge of [forces] of disbelief of infidels, Jews, crusaders, Americans and Britons ... We will scare you with the help of Allah... Who are you, O descendants of pigs and apes ... Bush [you] little dwarf."

Mr. Al-Samaraai added: "Jihad is an obligation of every individual Muslim ... O [sons] of the Iraqi people, you are the warriors of jihad...I pray to Allah at this blessed hour ... that he turns the month of Ramadan into a month of victory... O Allah, let the infidels fight each other, and dry their blood in their veins... shake the ground under their feet ... 'Allah Akbar' to America ... Allah will settle his account with you, Bush!"

In his Friday sermon broadcast on Palestinian Authority Television on October 31, 2003, Sheikh Ibrahim Mudeiris discussed Ramadan: "Has the time not come for Muslim servants of Allah to say, 'We will establish our state in Ramadan' ... 'We will achieve victories in Ramadan as did our predecessors' ... 'and liberate the blessed Al-Aqsa [Mosque] in the month of Ramadan?' It is surprising that the soul of a Muslim ... fasts during the day and prays and reads the Koran through the night does not call on him to fight the battle of jihad to liberate the blessed Al-Aqsa Mosque and the land of Palestine."

Saudi cleric Hisham Khreisyat appeared on Saudi Al-Majd TV last year during Ramadan on October 17, 2004, explaining: "This great month [of Ramadan] is a month of fasting, a month of blessings and victory, a month of jihad and martyrdom. Dear viewers ... [Ramadan] has always been the month of jihad ... to this day. They have always been battles of victory and liberation ... a month of martyrdom and martyrs. Thus we have been taught by history, by the Koran, and by this living nation who, Allah willing, will remain alive thanks to the blood of its martyrs, the sweat of its prisoners, the blood of its wounded, and the sacrifice of its sons."

Following terrorist attacks in Europe this year, the issue of incitement to violence by Muslims in the West is currently the subject of debate this Ramadan. According to Iran's Islamic Republic News Agency, on September 30, 2005, England's "Ramadan Radio" stations which for the past 13 years have had many of that country's estimated 1.8 million Muslims tuning in while they broke their daily fast, have been warned by the government "not to include material which condones or glamorizes violent behavior." The administrator for Radio Ramadan Birmingham, Ejaz Siddique, was quoted stating defiantly that he was "not going to shy away from talking about jihad."

Al-Qaeda's Ramadan Greetings: 'This Is Your Festive Season'

October 12, 2005

Every year since September 11, 2001, there have been threats of potential Al-Qaeda attacks during the month of Ramadan. This year is no exception, as there have been reports of an all-encompassing "Great Ramadan Offensive," warnings of potential attacks on the subway system in New York, and an attack in Bali aimed at Western tourists. Each year also has seen Al-Qaeda release communiques associated with jihad during this holiday.

In Ramadan 2004, the 27th issue of Sawt Al-Jihad, an Internet journal affiliated with Al-Qaeda in Saudi Arabia, included articles devoted to jihad during Ramadan. The opening editorial stated: "Muslims! Go out to [fight] Jihad for the sake of Allah! Paradise has already flung open its gates and the virgins of paradise are already decked out in anticipation of their grooms ... You should at the very least pray for your brothers the Jihad fighters, who gave their lives ... in revenge for the pride of the Muslim nation that was wounded by the cross bearers and ... Jews."

In reference to bombings in Taba, Egypt, that killed 32 Israelis, the writer, Sheikh Aamer Bin Abdallah Al-Aamer stated: "We pray for the Jihad fighters [who] ... attacked a hotel in which Jews were gathered and killed dozens of them as an offering to Allah before Ramadan. We ask of Allah that he not deny them [their] reward ..." It concluded: "This month of Jihad has come with all its blessings ... Come closer to Allah through the blood of infidels, do not relent in spilling [their blood], and through [this blood] wipe out humiliation and disgrace from among your Muslim nation! Make this month like the month of the Battle of Badr, the conquest of Mecca, [the conquest] of Shaqhab, and other Islamic victories. We ask Allah to turn this Ramadan into a month of glory, victory, and might, to hoist high in [this month] the banner of religion, to strengthen Islam and the Muslims, to humiliate polytheism and polytheists, to wave the banner of monotheism, to firmly plant the banner of Jihad, and to smite the perverts ..."

Sheikh Aamer Bin Abdallah Al-Aamer wrote another editorial in the Al-Qaeda journal stating: "Men of Jihad, this is your festive season since Jihad, in a state of fasting, has a particularly delectable taste for the believers ... wonderful it is to delight in the breaking of the fast and to taste the killing of infidels, to delight in the sound of the wailing of tyrants and lowly degenerates ... [This is a] month of fighting, of self-sacrifice, of military victories, of the victory of religion and the humiliation of the infidels. The most courageous and finest forays took place in Ramadan ..."

In November 2003 during Ramadan, a previously unknown Al-Qaeda group, Islamic Bayan Movement, issued a communique that opened with a passage from the Koran, then said: "The American rulers, who think [of] themselves as good-doers. Allah is the witness that they write the end of America by their own hands ..." The communique further warned: "Our Muslim brothers in America, we ask you to immediately leave the following cities: Washington, D.C., New York, and Los Angeles."

Another communique, by Abu Shihab Al-Qandahari, threatened Americans with a terror attack on the Muslim holiday of Eid al-Fitr in December 2002. Koran verses cited in the communique that were addressed to Americans: "You did not understand the reasons for the raids of Washington and New York ... This structure of theirs, which they have built, will ever continue to be a source of disquiet in their hearts, unless their hearts are cut to pieces and they perish ... Allah's words were rightly said about those who slew you on that white day: 'Allah has purchased from the believers their souls and their property in return for the promise that they shall have Paradise, for they fight in the cause of Allah and they slay and are themselves slain. This is a promise that He has made incumbent upon Himself as set out in the Torah, and the Gospel, and the Koran ..."

The communique added, "Allah promises [the Muslims] Paradise, and to you he has promised defeat ... Oh community of Americans, we repeat for the thousandth time: leave our lands and stop supporting the Jews, otherwise men faithful to their vow to Allah will come upon you [to martyr themselves] ..."

It should be expected that during the rest of Ramadan this year, the main Islamist sermons and jihadi Web sites will continue to incite attacks, particularly against Americans.

The Darfur Conspiracy

October 19, 2005

With the backing of the Arab dominated government of Sudan, nomadic Arab militias have been committing genocide against the mainly Muslim tribes people of Darfur since 2003. An editorial in the Washington Post on August 12, 2005, titled "Arabian Shame," criticized wealthy governments throughout the world for not properly giving aid to Darfur - while America had given 53% of all donations.

The editorial was especially critical of oil-rich Arab countries: "This Arab indifference is shameful. The victims of Sudan's worst crisis, in Darfur, are Muslim, and aid to non-Muslim southern Sudan is essential. ... Arabs have every reason to care about Sudan, and yet they have done far less than remote non-Muslim countries."

An article titled "The Arab Silence on Darfur Revisited" by Abu Khawla, the former chair of the Tunisian section of Amnesty International, similar ly captured the position of many Arab reformists. The article, which appeared on the liberal Arabic Web site, Middle East Transparent, on December 22, 2004, stated, "The catastrophe unfolding these days in Darfur ... is considered to be the worst humanitarian crisis in the world. ... [Secretary-] General Kofi Anan described the matter as a collective massacre of civilians. ... In contrast, a deafening silence was observed throughout the Arab world on the horrendous crime being committed by their fellow Arabs in Sudan..."

The former editor of the London Arabic daily Alsharq Al-Awsat, Abd al-Rahman al-Rashed, published an op-ed on June 24, 2004, about the Arab press's indifference to Darfur: "They are not the victims of Israeli or American aggression; therefore, they are not an issue for concern. This is how an approach of indifference toward others outside the circle of conflict with foreigners, and of permitting their murder, is spread as you read and write about the Darfur crisis ... Is the life of 1,000 people in western Sudan less valuable, or is a single killed Palestinian or Iraqi of greater importance, merely because the enemy is Israeli or American? ...As for Arab intellectuals ... who consider any blood not spilled in conflicts with foreigners to be cheap and its spilling justifiable - they are intellectual accomplices in the crime..."

While some Arab writers have denounced the Arab indifference to Darfur, others have spread conspiracies surrounding it. On the Saudi government TV Channel 1, on February 3, 2005, Saudi journalist Suheila Hammad stated: "By Allah, this is a conspiracy. [The Americans] say so... There is a conspiracy in Sudan - Sudan is being divided so that Darfur will become a secular state, independent from Sudan, the south will become a Christian state..."

It should not be surprising that the Iranian press has also spread conspiracies about Darfur. In the Tehran Times of July 13, 2005, Hassan Hanzadeh, wrote: "The war in southern Sudan and the Darfur crisis have caused serious economic and political problems ... In the region, neighboring countries, with the help of the Zionist regime, which is trying to weaken African Muslim countries by triggering civil wars, tried to dismember the great African Islamic country of Sudan by arming the Sudanese rebels. Their main objective is to create a Christian country on the banks of the Nile in order to end the domination of Egypt and Sudan over the world's longest river."

Shortly before the last presidential election in America, some influential Arabs said Darfur was actually a conspiracy to help President Bush win. The Sudanese ambassador to Cairo, Dr. Ahmad Abd Al-Halim, spoke on Saudi Al-Majd TV on August 11, 2004, and said "extremist" voters in America want to turn Sudan into a Christian state.

The Egyptian government controlled Al-Ahram Al-Arabi published an extensive investigative report by Dr. Amani Al-Tawil titled: "The Key to the American Voting Booths Is in Darfur: The Plot Which Is Called Oil" on June 31, 2004. Similarly, an editorial in Egypt's Al-Ahram Weekly on June 29, 2004, questioned if "the Sudanese government is undertaking operations of ethnic cleansing against the inhabitants of Darfur, and especially against non-Arab tribes." The article added: "The suspicion in the Arab world is that America's eagerness to intervene in Darfur is an American conspiracy to gain control of Sudanese oil."

While this and other editorials question if genocide is occurring in Darfur, just over the past couple of weeks killings have continued. One typical attack occurred on September 28, when 400 Arab fighters on camels and horseback attacked the Aro Sharow refugee camp killing 34 civilians, as Sudanese government helicopters flew overhead. The Arab press mainly ignored the attacks - truly shameful indeed.

Coming to America: Saudi Hate TV

October 26, 2005

An article in the Saudi Gazette of June 21 titled "Saudi Suicide Bomber was a Comedian" reported that one of its on-air employees was martyred in a terror attack in Iraq: "Al-Majd channels acknowledged this week that Saudi suicide bomber Mohammed Shathaf Al-Shehri, who attacked the Al-Hamra Oasis Village compound in Riyadh on May 12, 2003, was a math teacher and a hired comedian for the television station."

It further stated, "Just as the Ministry of Education was shocked to be informed that one of their own mathematics teachers was a suicide bomber and just as King Saud University was also shocked that one of its former students and associates was a suicide bomber, Al-Majd channel also is shocked that one of its freelance comedians was a suicide bomber."

Responding to claims that Al-Majd is a "hatching ground for terrorists," a statement by the network defending the channel's integrity as reported in the Saudi Gazette explained: "The programs on the TV screen will speak out our innocence."

Yet many influential Saudi religious figures frequently appear on the channel calling for jihad against the U.S. Saudi law professor Saleh bin Muhammad Al-Sultan was interviewed on March 6, 2005, discussing how Muslims should view the events in Iraq from the perspective of Islamic law and why it is correct "to wage jihad against the occupation."

He added: "The U.S. itself admits it is occupying Iraq. If it is an occupier, the people of Iraq have an individual duty to fend off this occupier ... Therefore, the jihad of the Iraqi people must target this occupier."

On December 21, 2004, Saudi professor Abd Al-Aziz Al-Fawzan explained: "Jihad is an individual duty applying to the Iraqi people. They need to wage this jihad against this enemy until it leaves their country."

The channel is incredibly anti-American. One of many examples includes Saudi professor Nasser Bin Suleiman Al-Omar, who appeared on a June 13 program.

Mr. Al-Omar, who runs a popular Islamic Web site, www.almoslim.net, discussed the approaching collapse of America, aided by the Muslim-American community: "America is collapsing from within ... Islam is advancing according to a steady plan, to the point that tens of thousands of Muslims have joined the American army and Islam is the second largest religion in America. Today, America is defeated. I have no doubt, not even for a minute, that America is on its way to destruction."

The Saudi Gazette of June 21 provided details about the programming of Al-Majd: "Saudi-owned and marketed to be an Islamic alternative to media broadcasting. The channels are very popular among religious Saudis and Arab Muslims. They appeal to many religious Muslims because they do not broadcast music or uncovered women."

One Islamic program included an interview by Saudi cleric Saleh Al-Munajd on October 25, 2004. The sheikh explained that globalization really meant converting the world to Islam: "It is our duty to globalize the world around Islam and call the world to join Islam: 'You may warn the Mother of Cities and those around her.' This means the entire world. This religion was sent to the entire world, globalization, in principle."

Another religious program on Al-Majd included an interview with Saudi professor Walid Al-Shuzi from Al-Riyadh University on March 18. The sheikh explained that Islam teaches Muslims that they have a religious duty to hate Jews and Christians: "Some people say that when you leave a mosque you must love the Jews and Christians ... We are forbidden to attack them, but we have a duty to hate them, as is written in the Koran and the Sunna."

Following last year's tsunami, many notable Saudi religious figures appeared on Al-Majd with anti-Christian comments. Saudi Sheikh Fawzan Al-Fawzan, who taught at the Saudi Supreme Court of Justice, said on December 31 that Allah's destruction of these countries was timed to coincide with Christmas - a time of rampant fornication and perversion.

It should be noted that after NBC aired statements from his Al-Majd appearance, which were provided by MEMRI TV, the London Arabic daily Al-Hayat interviewed the director general of Al-Majd. He confessed that Mr. Al-Fawzan's statement was "met with great displeasure in American circles."

Following other channels already available in America, including Saudi TV 1 and 2, as well as Iqra TV, a January 26 article on Magnaquest.com reported that Al-Majd is in the process of trying to expand to America.

To view segments of Al-Majd TV, visit www.memritv.org.

‘A World Without America’

November 2, 2005

President Ahmadinejad’s speech to the Iranian student organization Union of Islamic Students Association’s “World Without Zionism” conference last week stating, “Very soon, this stain of disgrace [i.e. Israel] will vanish from the center of the Islamic world” has been widely condemned internationally.

In Iran, however, his speech was surrounded by celebrations and proclamations - which were part of Al-Quds Day. This included a reception for ambassadors from Muslim countries “thrown in their honor” by Iran’s Foreign Ministry on October 27.

A legal adviser to the Palestinian Authority on “peace negotiations,” Diana Butto, was pictured by the Islamic Republic News Agency on October 28 speaking to an audience in Iran about “her support of the intifada.” A Hamas leader, Khaled Meshal, issued a statement to the conference with a warning to America: “The Islamic nation feels ready to confront Zionism and the United States.”

As the Mehr News Agency reported on October 20, programs to celebrate a “World Without Zionism” and Al-Quds Day were also planned for abroad. These included events hosted by Iranian embassies throughout Europe. For example, Iran’s ambassador to Spain, Morteza Alviri, hosted a ceremony for ambassadors of Islamic states at his office.

The Union of Islamic Students Association conference has many initiatives going on as part of official “World Without Zionism” celebrations. One includes a global competition on cartooning, painting, and graphic design for children between the ages of 7-18. Thirteen cartoons have already been selected as winners from countries such as America and the Netherlands.

Another competition focuses on “A World Without America.” As an IRNA headline from October 27 quoting Mr. Ahmadinejad proclaimed: “World Without U.S. ... Achievable,” he explained. “They [ask]: ‘Is it possible for us to witness a world without America and Zionism?’ But you had best know that this slogan and this goal are attainable, and surely can be achieved.”

In an analogy between the former Soviet Union and America, he added, “Nobody believed that we would one day witness the collapse of the Eastern

Imperialism [i.e. the U.S.S.R]... it was an iron regime. But in our short lifetime we have witnessed how this regime collapsed."

"A World Without America" was also the main focus of the Union of Islamic Students Association's conference last year. Many prominent Iranians participated such as Ali Larijani, Iran's Supreme Council for National Security. He was quoted on the Iranian News Channel on November 8, 2004, stating: "As the nations (of the world) reach intellectual growth, political, and economic independence, they will thwart America."

The deputy chief of the Revolutionary Guards, Mohammad Baqer Zol Qadr, also appeared, stating: "When we talk about the 'World Without America,' we mean a world governed by peace and justice. ... Unfortunately, today America is the symbol of these deficiencies and distortions."

An official video, which can be seen on www.memritv.org, with footage from the attacks of September 11, 2001, when both planes hit the World Trade Center, was shown at the conference. Classical music played as another image appeared with the word "America" spelled out in flames. Each letter disappeared - leaving an empty screen.

As the video continued, the music grew louder and the words "The World Without America" appeared in English and Farsi - accompanied by a picture of the Statue of Liberty. Lady Liberty's body had a skull imposed on it.

Ayatollah Khomeini could be heard in the background: "Oh sleepy ones, oh distracted ones, wake up and look around you: You have built your house near wolves' dens. [A picture of President Bush is shown.] Rise, this is no place to sleep. We must shout: Rise up soon because the world is not safe from the hunter. America is lurking for you, and will not give up until it destroys you completely. Indeed, if there was a global mobilization of Muslims, would anyone dare to be so impudent and act so cruelly toward the spiritual offspring of the Messenger of Allah?"

Al-Jazeera in Plain English

November 9, 2005

Al-Jazeera's international English language news service will be previewed at the Dubai-based Media and Marketing Show 2005 later this week. The new channel has made headlines with high-profile hirings intended to make it more palatable to viewers and advertisers in the West. The team includes a veteran BBC correspondent, David Frost; a former American military spokesman, Josh Rushing, and a former CNN anchor, Riz Khan.

On August 4, 2004, this column detailed the arguments over Al-Jazeera's role in inciting its viewers to violence. In addition to receiving alleged tips on impending terrorist attacks (in order to tape them for broadcast); the channel continues to highlight graphic terrorist activities, such as beheadings, on the air. Of late, Al-Jazeera has repeatedly shown killings of American soldiers in Iraq, almost on a daily basis.

Last summer's column also noted that Al-Jazeera correspondents have been arrested worldwide for connections to terrorist activities. A notable example is Tayseer Alouni who was first arrested by Spanish authorities on September 5, 2003, for "furnishing Al-Qaeda with funds." Mr. Alouni interviewed Osama bin Laden shortly after September 11, 2001, and just weeks ago received a seven-year jail sentence in Spain for alleged collaboration with Al-Qaeda.

It should be noted that the second "Sout Al-Khilafa," Al-Qaeda's Internet news broadcast, from September 28, 2005, included a segment on Mr. Alouni. The Al-Qaeda anchor urged viewers to mobilize public opinion in favor of the jailed journalist.

The anchor stated: "According to the breaking news ... the Spanish court deliberating the case of our Muslim brother, Tayseer Alouni, sentenced him to seven years in prison. The Sout Al-Khilafa media company released a communique over the Internet, harshly condemning the conduct of the Spanish Crusader infidel court towards the Muslim journalist Tayseer Alouni ... The communique expressed complete solidarity with our brother Tayseer Alouni in this crisis and in light of the injustice from which he is suffering because of his reliable and completely objective reporting, which complied with the internationally accepted journalistic standards."

Al-Jazeera has been under scrutiny by American government officials including Defense Secretary Rumsfeld, who has accused it of promoting terrorism on multiple occasions. A former ABC News producer who has worked for both Bush administrations, J. Dorrance Smith, is currently up for the chief Pentagon spokesman position. On April 25, 2005, he wrote an op-ed in the Wall Street Journal titled, "The Enemy on Our Airwaves," which criticized American networks for rushing to air news segments from Al-Jazeera, saying "this is a powerful tool for the terrorists."

Mr. Smith also brought up many questions about "the precise terms" of the relationship between Al-Jazeera and terrorists: "When the terrorists want to broadcast something that helps their cause, they have immediate and reliable access to Al-Jazeera. This relationship - in a time of war - raises some important questions: What does Al-Jazeera promise the terrorist organizations in order to get consistent access to their video? Does it pay for material? Is it promised safety and protection if it continues to air unedited tapes? (No Al-Jazeera employee has been killed or taken hostage by the terrorists) ... Does Al-Jazeera promise the terrorists that it won't reveal their whereabouts and techniques as a quid pro quo for doing business?"

While the debate surrounding the reported connections of Al-Jazeera to terrorism rages, the network continues to show objectionable programming. One recent example was a special called "The Truth Behind 9/11." Part Four appeared on September 30, 2005, and was devoted to detailing the Mossad's involvement in the attacks of September 11, 2001. One person interviewed was the editor in chief of the London Arabic daily Al-Quds Al-Arabi, Abd Al-Bari Atwan.

As MEMRI documented in an Al-Jazeera special two months before September 11, Mr. Atwan was among those who appeared justifying Osama bin Laden's actions at that time, and claimed that Mr. bin Laden has no intention of destroying America.

The coming English Al-Jazeera network should be expected to continue airing such programs.

Syrian Conspiracies

November 23, 2005

This fall, the streets of Syria have been abuzz with ongoing conspiracy theories to account for the country's woes. President al-Assad's speech two weeks ago warned against conspiracies in the form of "psychological warfare" and "black propaganda" with Israel as the main beneficiary. Other usual suspects - Jews, Zionists, and America - are also blamed for the ongoing terrorist activities in Syria and Lebanon.

The most far-reaching conspiracy theory surrounds the assassination of the former Lebanese prime minister, Rafik Hariri. The Syrian defense minister, Hassan Turkemani, was quoted shortly after the attack in the United Arab Emirates paper Al-Bayan on February 17, stating: "The only one to benefit is Israel." In an article in Al-Khaleej on February 17, Wisam Hamoud, a political commentator, accused Israel and America of perpetrating the crime: "After assassinating al-Hariri, the Israelis and Americans can claim that there is a worsening in the region's stability, and that there is a need for increasing democracy and for a new occupation. Thus, al-Hariri's assassination, like many murders in Iraq, is an act of Israel's secret service apparatus, or of America's mercenaries hired to murder enlightened progressives in the Arab world."

Following the assassination, the United Nations announced it would establish an official investigation headed by Detlev Mehlis. Not surprisingly, the Syrian press has attacked him and his investigation. Writing in Al-Thawra on October 9, Lebanese journalist Maria Maloof called him "the German fox," comparing him to "Kojak, Columbo, and Inspector Gadget."

An unnamed "Syrian citizen" discussed the Mehlis report in Teshreen on October 25: "The report was prepared to please the U.S. administration, Israel, and their allies. The means used to assassinate the late Hariri were very sophisticated. The technology is only available to a superpower developed secret services like the CIA or the Israeli Mossad."

On October 31, the Syrian parliament had official proceedings on the Mehlis report. One unnamed Syrian lawmaker said: "You should look for the murderers of Hariri in Tel Aviv and Washington." Lawmaker Huneim Namar said: "The greatest thing that the Americans and Israelis achieved from this Mehlis report is to divert attention away from any possible role played by the Israeli Mossad, [and] the American CIA, as well as to direct the spotlight exclusively on Syria." Lawmaker Anwar Ubeid added: "What is happening today is an indication that America and Bush are coveting this nation's resources. Syria is the only thorn to remain in the eye of Zionism ... the Mehlis report is a clear attempt to pressure and harm Syria.'

Strongly hinting at America and Israel, Syria's minister of expatriates, Bouthaina Shaaban, wrote in the London Arabic daily Asharq Al-Awsat on October 11 about Syria's "enemies" scheming against the country: "How do Arabs feel when they read on the Internet that the late Rafik Hariri was assassinated for ... his stand against the disarmament of Hezbollah and the international campaign he led regarding this issue and his rejection of a large U.S. military base in Beirut." Mr. Shaaban, who frequently visits America to meet with Arab-American leaders, explained the Mehlis report is an attempt "to wipe out Arab ideology" and concluded by asking, "Who will be the main beneficiary from the series of events planned and executed by our enemies?"

The most shocking conspiracies came from the Syrian government press that has blamed Hariri's sons for his assassination. The editor of Al-Thawra, Fayez Al-Sayegh, wrote on October 16: "The international investigation ... ignored two factors, namely Israel ... and al-Hariri's heirs - his sons, his associates, or others who gained from his death." An October 9 article in the Syrian weekly Al-Igtisadiyya also named Sad al-Hariri as responsible for a number of bombings and assassinations that occurred in Lebanon in the last few months.

A 12-page article in Teshreen on November 11 titled "Will the Real Killer of Hariri Please Show His Face" explained that as the December 15 deadline for the Mehlis report to be submitted to the United Nations approaches, the real culprit, the Zionists who "control the planet" with ruthless plans to destroy tiny Syria, should be held accountable.

It should be noted that the Syrian regime's attempt to blame its current problems on others has had little success in easing the pressures it is facing. Some people are now openly questioning how much longer the al-Assad dynasty will continue.

George Galloway in the Middle East

November 30, 2005

Two hundred million barrels of oil is what British lawmaker George Galloway received from Saddam Hussein "for misleading public opinion concerning the crimes perpetrated by Saddam against his people." This is the claim of Iraqi writer Fadhil Rashad Al-Khafaji in a November 4 article on the Web site Al-Jiran.

In a bold statement calling for Mr. Galloway "to be tried," Mr. Al-Khafaji wrote: "The Iraq people will never turn a blind eye to those who robbed them. You [Galloway] robbed them." Mr. Al-Khafaji also noted that Saddam's payments went to individuals such as Mr. Galloway, who "continue to fiercely defend Saddam when they are interviewed on satellite TV - and this despite the evidence of his crimes..."

This year Mr. Galloway has visited the Middle East numerous times, appearing on Arab TV, mainly to attack America and Britain, while rallying in support of Syria's president, Bashar al-Assad. On November 13, he appeared at a pep rally at Damascus University in which he compared the Iraqi resistance fighting America to the British fight against Nazi Germany, as well as praising the "hero Hugo Chavez" for standing up to President Bush.

In another summer visit to Damascus, Mr. Galloway gave a July 31 interview to Syrian TV in which he called on the countries under Messrs. Bush, Blair, and Berlusconi to be "resisted." Mr. Galloway also directed a statement to the Arab world: "Two of your beautiful daughters are in the hands of foreigners - Jerusalem and Baghdad. The foreigners are doing to your daughters as they will. The daughters are crying for help, and the Arab world is silent. And some of them are collaborating with the rape of these two beautiful Arab daughters."

The same day Mr. Galloway appeared on Al Jazeera stating: "It's not the Muslims who are the terrorists. The biggest terrorists are Bush, and Blair. ... Bush doesn't have any civilization. ... We believe in the Prophets, peace be upon them. He believes in the profits, and how to get a piece of them. That's his god." In support of the terrorists in Iraq, Mr. Galloway explained, "The Iraqi resistance is ... defending all the Arabs, and they are defending all the people of the world from American hegemony."

Mr. Galloway also gave an interview to the editor of the leading Syrian government daily Teshreen on July 8, which, as of this writing, can be found on the front page of its Web site. After criticizing the West for calling Hamas and Hezbollah "terrorists," he was asked his opinion on why the press is unfair to such organizations: "They are distorting facts and launching disinformation campaigns against the Arabs and Muslims and freedom fighters in general. Mr. al-Assad said in a speech the Iraqi people are resisting occupation troops. And that we have to differentiate between those who have destructive weapons and practice all forms of terror and violence against peoples and those who have nothing to defend themselves with except to carry a knife, a stone or to carry a bomb and take it to the positions of the occupiers and explode it."

Mr. Galloway's interviews were not well received in London. In an article on August 4, the Sun called him an "Evil Traitor." Former Labor lawmaker and ex-serviceman Eric Moonman called on the speaker of the House of Commons to probe his conduct for "putting at the risk the soldiers who served the country he is supposed to represent." The Evening Standard editorial on August 5 added: "Taking a stand against the war in Iraq is absolutely legitimate; justifying the inciting the killings of British troops and Iraqi policeman is not. Trouble is, what to do with Mr. Galloway? Under the Home Secretary's new rules, it is to be easier to deport preachers of terrorism. But where do you deport Mr. Galloway? Scotland?"

Mr. Galloway ended his Teshreen interview bragging about his appearance before the Senate this summer: "The American senators were not expecting their old crimes to be exposed on live TV ... I made a lot of people happy ... and I made the worst people in the world sad. And this is a good day's work."

The last laugh may yet go to Mr. Galloway's critics. Iraqis including Fadhil Rashad Al-Khafaji are now calling for him to be put on trial. The Department of Justice announced in late October that it will investigate if he lied to the Senate. The Charities Commission in England and Wales, as well as the House of Commons Parliamentary Commission for Standards are reportedly now also investigating Mr. Galloway's actions.

Miss Al-Qaeda

December 14, 2005

The Saudi daily Al-Watan reported November 27 that 122 women were recently arrested in Iraq for "attempting to carry out suicide operations." On December 6 two women bombers were successful in detonating themselves and 43 Iraqis.

Other women lining up to become Al-Qaeda suicide bombers include Iraqi Sajidi Mubarak Atrous al-Rishaqi, whose failed attempt last month in the Amman attacks and subsequent confession on Jordanian TV were seen throughout the world. Another notable example was Muriel Degauge, the Belgian convert who blew herself up in an attack on an American patrol in Iraq last month.

The successful campaign by Al-Qaeda to enlist female suicide bombers began with the launch of Al-Khansaa, Al-Qaeda's women's magazine last year. The magazine was named after Al-Khansaa bint Omar, a poetess of the pre-Islamic period who converted to Islam during the time of Muhammad, and is considered to be the "mother of the Shahids [martyrs]." When her four sons died in the battle of Al-Qadissiyya, she did not mourn but thanked Allah for honoring her with their death. Following suicide attacks in Israel, the mothers of the bombers are often found making similar statements on Arab TV.

An editorial in Al-Khansaa's first issue in August 2004 stated: "Our goal is Shahada [martyrdom] for the sake of Allah, and our goal is [to gain] the pleasure of Allah and His Paradise. We stand shoulder to shoulder with our men. ... We will stand covered by our veils and wrapped in our robes, weapons in hand, our children in our laps, with the Koran and the Sunna of the Prophet of Allah directing and guiding us. The blood of our husbands and the body parts of our children are the sacrifice by means of which we draw closer to Allah.... The wind of Paradise is already blowing."

Another article by Umm Badr titled, "Obstacles in the Path of the Jihad Warrior Women," details how women must prepare for jihad, including emotionally, militarily, and physically to prepare for "direct conflict with the enemy."

A November 25 report in the London Arabic daily Asharq Al-Awsat dis cussed Web sites that are affiliated and sympathize with Al-Qaeda which often refer to the role of women in Chechnya's jihad. The article stated that Chechen rebel leader Shamol Basayev always took pride in the "Black Widows" who engaged in suicide operations against the Russians after losing their husbands in battles. The widows wore black gowns that concealed explosive belts and played an important part in the hostage taking operations of the Moscow theater and Beslan school massacres. Approximately 27 women successfully carried out suicide attacks against the Russians.

The article also quoted London Islamist Hani Al-Sibai, who heads the Al-Maqrizi Center for Historical Studies. He claimed that Ms. Al-Rishaqi's failed attack was Al-Qaeda's "first attempted operation to be carried out by a women." In Iraq he explained, "Women do not even need to seek permission from their husband, father, or brother to participate in jihad" and "that al-Zarqawi's group would not deprive women of the rewards of martyrdom if they are able to fight."

While "72 Black-Eyed Virgins," are the well-known rewards of male suicide bombers, one would-be female Palestinian bomber caught in Israel recently told the press that religious authorities taught her that the reward of women martyrs is to "become the purest and most beautiful form of angels at the highest level possible in heaven."

It should be mentioned that the most influential Arab religious, political, and media personality have celebrated female suicide bombers over the past few years. For example, the first female Palestinian suicide bomber, Wafa Idris, was called the Arab world's "Mona Lisa" and "Joan of Arc."

The Egyptian Nasserite weekly Al-Arabi, stated: "Wafa Idris... became the most beautiful of the women in this world. ... What is more beautiful than the transformation of a person from a chunk of flesh and blood to illuminating purity and a spirit that cuts across generations?"

Salah Muntasir wrote in the leading Egyptian government daily Al-Ahram: "Wafa Idris has not faded from my [mind] since I saw her picture for the first time. ... Her dreamy eyes and the mysterious smile on her lips, that competes with the famous smile some artist drew on the lips of Mona Lisa - Wafa's smile is more beautiful."

Some Arab writers have denounced female suicide bombers, but overall their voices have been drowned out by those who support them.

A Saudi Prince's Indecent Proposal

December 21, 2005

Even though he is one of the richest men in the world, as the recently released biography of billionaire Saudi Prince Alwaleed bin Talal details, his name first made headlines only after September 11, 2001, when Mayor Giuliani rejected his $10 million proposal after what the prince said about the root causes of the terrorist attacks in America.

The biography quotes directly from Mr. Giuliani's book "Leadership," which recalled the episode: "When the Prince arrived, he was wearing an opulent gold robe and headdress... He gave me a cashier's check for $10 million, for the Twin Towers Fund. Looking at the site from the small podium, the Prince was saying the right things... But something wasn't quite right. There was a smirk on his face, which seemed to carry over to his entourage. He was the only visitor who was unmoved by what he saw."

Prince Alwaleed's biographer, Riz Khan (anchor for the upcoming English Al-Jazeera news channel), added, "It was politics that angered many in the Middle East, who considered Giuliani to be bowing to pressure from certain groups ... One columnist in Al-Riyadh, a Saudi newspaper, wrote: [Giuliani] sacrificed the public interest for a private interest, manifested in his desire to draw closer to the Jewish electorate."

This text was actually from a October 11, 2001, MEMRI dispatch by Mahmoud bin Abd Al-Ghani Sabbagh. The book omits the article's headline, "Al-Waleed's Check, The Homosexual Governor [sic] & The Propaganda War," as well as insulting text from the article: "The words of [Prince Alwaleed] did not, of course, please the Jewish lobby ... Because the governor [sic] of the Big Apple is a Jew, he refused [to accept the donation] ... Giuliani said: 'The Prince's declarations are grievous and irresponsible; these Arabs have lost the right to dictate [to us what to do]. What we [America] must do is kill 6,000 innocent people. By Allah, I am amazed at your act, you Jew; everything Prince Al-Waleed said was true..." Mr. Al-Waleed's defender went on to attack New York City and Mr. Giuliani, calling him "a homosexual in a city in which dance clubs, prostitution, homosexuality, and stripping proliferate."

This month, the prince made headlines again when it was announced he donated $20 million to Harvard and Georgetown universities. The funds will be used at Georgetown to expand the Center for Muslim-Christian Understanding that will be renamed the Prince Alwaleed bin Talal Center. The prince told the Arab News on December 14 why he gave the money: "We are determined to build a bridge between Islam and Christianity for tolerance that transcends cultural and geographical boundaries."

Prince Alwaleed made billions in profits from businesses such as the ART TV network, which includes Iqra TV. According to his biography, the prince owned 30% of ART "and became fairly active in overseeing parts of it" but has sold some of its ownership.

A November 28, letter from Rep. Steve Rothman, a Democrat of New Jersey, and 22 other members of Congress to the new Saudi ambassador, Prince Turki Al-Faisal, was critical of Prince Alwaleed, his ownership of ART, and more specifically incitement on Iqra. Excerpts from the letter include: "We are greatly concerned about the ongoing incitement against Americans and Jews that is present in the programming content of Iqra TV, part of the ART TV network, which is partially owned by Saudi Prince Alwaleed bin Talal. The content of Iqra TV, which can be viewed in the United States and throughout the world, promotes anti-Americanism and anti-Semitism, incites its viewers to rampant Jihad, and advocates violence against Israelis. There is no excuse for spreading such hatred ... we strongly believe that the sponsorship of a television station that promotes Jihad and antipathy towards Americans and Jews by a member of the Royal Family must immediately cease."

After receiving the $20 million donation, Georgetown's president explained to the Arab News that the Alwaleed Center will help facilitate cultural and inter-religious dialogue. Yet programming on ART's Iqra TV includes Islamic religious figures calling on Muslims to confront non-Muslims; calls for the entire world to follow Islam, and for Christians and Jews to burn in hell for being nonbelievers.

Other notable examples include the Saudi deputy minister of religious endowments, Abd Al-Rahman Al-Matroudi, defining terrorism according to the Saudi government as: "Any act or statement that contradicts the Koran." More recently, on December 7, Iqra interviewed a toddler who discussed the upcoming "big battle" with the Jewish "dogs" and "villains."

Top 10 Conspiracy Theories: Part III

December 28, 2005

A Saudi journalist, Mshari Al-Zaydi, wrote about the "disease" of the Arab press blaming others for the Arab world's misfortunes in a London-based newspaper, Al-Sharq Al-Awsat, on November 20. "This huge obsession among some Arabs and Muslims regarding conspiracy theories and the belief that the world is lurking in wait to pounce on us, as if the world has no worries other than cooking up plans, policies, and moves in order to realize one objective only: to eliminate Islam, Muslims, and Arabs," Mr. Al-Zaydi wrote.

The following are the top 10 conspiracy theories of 2005:

10. An Iraqi Shiite imam, Jalal Al-Din Al-Saghir, gave a sermon on December 16 that was broadcast on Al-Furat TV. He called Al-Jazeera "a TV channel known to be guided by the Mossad and ... whose purpose is to damage ... Islamic interests."

9. On a program that aired September 30 on Hezbollah-backed Al-Manar TV, the director of Sweden's Radio Islam, Ahmad Rami, (who was found guilty of incitement against Jews and served time in prison) discussed Jews in the West who "have 100% complete control of the media, political parties, trade unions, and publishing houses."

8. Saudi Arabia's Al-Majd TV interviewed a Jordanian lecturer, Sheikh Ahmad Nawfal, about Jewish history on November 13. "David and Solomon were among our [Muslim] ranks. If Solomon had a temple, we would be worshipping Allah in it. We would not be worshipping idols and polytheism in it, like they do," he said. "There is no indication that a temple existed there."

7. Following the December 6 C-130 plane crash in Iran, the Iranian Student News Agency quoted the secretary-general of the Association of Muslim Journalists, Parviz Esmaeili, blaming, "U.S. state terrorism." Rambling about American

sanctions on Iran, Mr. Esmaeili called on, "all the heads of state in the world" to discuss at international meetings this "plane crash and similar incidents over the past two decades [in Iran] as evidence of the U.S. technological crimes."

6. Mr. Esmaeili went on to tell ISNA that if the attacks of September 11, 2001, were investigated, the probe would show that the perpetrators were "the U.S. government and not phantom players like Al-Qaeda." Other programs on Iranian TV devoted to the attacks included French author Thierry Meyssan's August 30 interview with Jaame-Jam2 TV and a November 17 interview with an Iranian filmmaker, Nader Talebzadeh, on IRINN TV.

5. Arab TV has also been rampant with September 11 conspiracies, including Al Jazeera's series "The Truth Behind 9-11," which was widely viewed throughout September, and a retired Egyptian general, Muhammad Khalaf, on Al-Mihwar TV on September 11 of this year detailing the American government's "secret plan" first developed in 1999 by "Bush senior."

4. Books on conspiracies continue to be popular throughout the Middle East. One devoted to the Bush family was the topic of a November 11 interview on Syrian TV featuring Syrian cleric Mohammad Said Ramadhan Al-Bouti: "Bush ... the grandfather of the current American president ... wrote a book about the life of the Prophet Mohammed. In this book, which was published in 1831, he says: 'As long as the Muslims' empire is not destroyed, God will not allow the return of the Jews to the homes of their fathers.'"

3. On December 14, the Syrian state owned daily Teshreen wrote about a book called "The Balance of Horror in an Open War," by Abdul Majid Ammar. It includes an analysis of the "Zionist entity and the conspiracies it hatched against the Arab nation in order to achieve its sinister expansionist plans."

2. Following the December 12 assassination of a Lebanese member of parliament and a leading anti-Syrian writer, Gibran Tueni, Iran's foreign minister, Hamid Reza Asefi, was quoted as saying it was "in line with ... the Zionist regime." A Teshreen headline on the story read, "Israeli Factor Should Not Be Dropped." On December 18, Al-Seyassah quoted Syria's information minister, Mahdi Dakhlallah, as saying, "Tueni was in debt and was killed by his debtors."

1. Many Arab writers have attacked Detlev Mehlis, the lead U.N. investigator of the assassination of a former Lebanese prime minister, Rafik Hariri. Mr. Mehlis's mother has been accused of being a prominent Zionist, as the Syrian

judicial Web site Al-Nazaha reported on November 17. The deputy editor of the Egyptian government daily Al-Gumhouriyya, Abd Al-Wahhab 'Adas, reported November 12 on his mother's "major role in bringing Jews from Germany to Palestine" and that "Mehlis's mother was killed on the Golan Heights by Syrian sniper fire."

2006

Must-See Hezbollah TV: Part I

January 4, 2006

According to its Web site, Al-Manar is a "Lebanese TV station," launched in 1991 that began broadcasting via satellite in 2000. The channel claims to focus on the tolerant values of Islam and "to promote the culture of dialogue and cooperation among the followers of the Heavenly religions and human civilizations." The Web site also states that "Al-Manar avoids cheap incitement" and presents a combination of religious programming, international and local news, sports, politics, culture, and children's shows.

The channel is officially associated with the Arab League and according to its Web site, "draws a large number of viewers inside the Arab world and in countries of immigration." The Lebanese Daily Star has reported that Al-Manar claims to reach an estimated 200 million viewers around the world, but the newspaper said a more realistic approximation is 10 million regular viewers. According to a French paper, Liberation, in Europe, North Africa, and the Middle East, Al-Manar reaches about 100 million homes.

The network first made headlines with the airing of the Syrian-produced Ramadan special Al-Shatat (Diaspora) in 2003. The series focused on the purported history of the Jews with classic anti-Semitic motifs including the blood libel. MEMRI was instrumental in bringing the issue of Al-Manar and Al-Shatat to the attention of world leaders including high-ranking American government officials; Germany's chancellor and its foreign minister; the Swedish and Australian prime ministers, and France's prime minister and president. After the French prime minister viewed segments from the series, he called for banning the channel in France, according to the French daily Le Monde.

Following the French ban of Al-Manar, Hezbollah has tried to create an impression that it is not connected to the channel. Yet Al-Manar's news director, Hassan Fadhlallah, is a Lebanese member of Parliament representing the Hezbollah party. In a 1995 interview with the Washington Post, an Al-Manar station manager said the station "belongs to Hezbollah culturally and politically."

Arab and Western press outlets also describe the channel as belonging to

Hezbollah. Writing in the Lebanese Daily Star on December 24, 2005, Rami Khouri called the channel "Hezbollah-owned Al-Manar." On November 21, 2005, Al-Jazeera TV reported: "The Al-Manar TV station, speaking in the name of Hezbollah," and on May 14, 2005, the Saudi daily Al-Riyadh wrote: "Al-Manar TV, which belongs to Hezbollah." Recent examples from Western sources include Le Monde on December 7, 2005: "Al-Manar - the Lebanese Shiite Hezbollah satellite TV" and Reuters, which on December 5, 2005 called Al-Manar a "mouthpiece of the ... Hezbollah group."

As Al-Manar's chairman, Nayyef Krayyem, who was also chief of Hezbollah's information department, told the Christian Science Monitor in 2001, "Al-Manar is an important weapon for us. ... It's a political weapon, social weapon, and cultural weapon." Hezbollah's deputy leader, Sheikh Naim Qassem, was quoted stating Al Manar's message is simple: "Jihad is the only way to salvation, and the experience of the Islamic and the Lebanese resistance is the best proof."

The channel is used to spread Hezbollah's message in Lebanon as well as throughout the world. In an interview with Islamonline in January 2001, an Al-Manar TV broadcaster, Fatima Bari, said Hezbollah's activities have "enriched Al-Manar TV through its Jihad, the blood of its martyrs, and everything it sacrificed" and that "Al-Manar TV waged an extremely harsh media war against the Zionist[s]."

In the recent Lebanese elections, Al-Manar got the terrorist organization's message out. No other groups or parties running for office had a TV station at their disposal. One Lebanese politician, 'Ali Trabulsi, appeared on the channel during the early round of voting on June 5 declaring, "I vote 'yes' for the resistance [i.e. Hezbollah]. I have written it with the blood of my 10 fingers." He then took his bloody finger and wrote on his election form: "Yes to the resistance." The result: His group won 14 out of 128 seats. and Hezbollah emerged as a power in the Lebanese Parliament.

In the coming weeks, this column will focus on how the channel is influencing the Arab and Muslim world in general and in particular spreading anti-Americanism and support for terrorism.

Must-See Hezbollah TV: Part II

January 11, 2006

One of the most authoritative books written on terrorist groups operating in America, Steven Emerson's "American Jihad," has asserted that Hezbollah is active in American cities, including Detroit, Philadelphia, New York, and Washington, D.C. Most notable was a Hezbollah cell operating in Charlotte, N.C. It was taken down following an FBI sting in July 2000 whereby 18 people were charged. According to the indictment, they were guilty of providing training, communication equipment, and explosives to Hezbollah, "in order to facilitate its violent attacks."

Hezbollah does maintain an elaborate network in America. Staff from its television network Al-Manar have lived and been educated here, including a translator, Mohammad Abdullah, who graduated from the University of Massachusetts. Al-Manar also reportedly has a Washington-based correspondent named Muhammad Dalbah.

Al-Manar's Web site, www.manartv.com, is registered in Seattle, Wash., by a company called eNom, with Internap Network Operations. Previously, the Web site was hosted in New Jersey. According to a May 31, 2005, report in the St. Petersburg Times, it was taken down temporarily but was up again a week later with another provider. Al-Manar Webmaster Mohammed Obeid explained: "Companies that do hosting for us are getting afraid of the consequences by the U.S."

Anti-American figures are occasionally interviewed on Al-Manar from Washington. For example, the editor in chief of the Washington-based Middle East Magazine, Ahmad Yusuf, appeared on December 30, 2004, and said Muslims were not involved in the attacks of September 11, 2001. He called them a grand scheme designed by Israelis and American right-wing forces, including "evangelical Christians." He also said the American government itself attacked Pearl Harbor as an excuse to enter War World II.

In its broader strategy to reach out to Americans, Al-Manar has been useful to Hezbollah. During the last two years, delegations from families of victims of the September 11 attacks, along with members of the Presbyterian Church, have appeared on the channel in meetings with the terror organization. In one instance, the deputy leader of Hezbollah, Sheikh Nabil Qauq, called President Bush's and America's "aggressive inclination a real danger to all monotheistic religions." A Presbyterian elder, Ronald Stone, stood at Mr. Qauq's side and said, "We treasure the precious words of Hezbollah and your expression of goodwill toward the American people."

In May 2004, Al-Manar invited foreign college students studying in Lebanon, from countries such as Australia, Russia, and America, to participate in a documentary in support of Hezbollah. Three American students took part, including an American University of Beirut graduate student, Stephanie Tournear. She was quoted in the Daily Star as saying, "It's a shame you can't state your opinion or observations regarding Hezbollah in the U.S." Another American student who would not be identified said, "I decided not to be involved in the documentary, as it could have security and employment implications for me upon return to the U.S."

In an important step in the war on terror, the State Department added Al-Manar to its "Terrorism Exclusion List" in December 2004 for incitement to terrorism. Among other things, the designation means that anyone working for or helping the network can be barred from America.

Yet Al-Manar maintains vocal Arab and Muslim-American supporters. Osama Siblani, publisher of Dearborn Michigan's Arab American News, which according to its Web site, "is the largest, oldest, and most respected Arab American newspaper in the United States" was quoted in the Washington Post as saying, "I disagree with the State Department that it [Al-Manar] incites violence. ... By that standard, they should shut Fox News for inciting against Muslims." Texas Muslims for Islamic Change issued a statement that it was "dismayed at this development and considers it to be part of the American government's assault on constitutional rights" adding, "to date we have not seen properly documented evidence brought forward that would support the State Department's claim that Al-Manar 'preaches violent and hatred' or 'serves to incite ... terrorist violence.'"

Since it has been put on the State Department's terror list, the station has continued to attack America, describing it as a "plague" with commentators calling for jihad against the country. As Hezbollah's leader, Sheikh Nasrallah, declared to a rally of thousands yelling "Death to America," covered live on Al-Manar in February 2005: "We consider the current administration an enemy of our [Islamic] nation. ... Our motto, which we are not afraid to repeat year after year is 'Death to America.'" Next week's column will focus on anti-American incitement on Al-Manar.

Must-See Hezbollah TV: Part III

January 18, 2006

The 1983 attack on American Marine barracks in Beirut that killed 241 was Hezbollah's coming-out party. To this day, the attack is lauded on its TV channel Al-Manar. A Hezbollah "poet," Atef Moussa, appeared on May 22, 2005, and said, "Who says we are afraid of war? ... Who can compare to the men of Hezbollah? ... These enemies [the American military] turned out to be as light as cardboard. Bush knows it. Beirut remains dangerous for the Marines. Our proof is here, they left in shame. Our people sail the seas of martyrdom."

In an anti-American speech mocking the American military on March 8, 2005, Hezbollah's leader, Sheikh Nasrallah, also referred to the attack: "I address the following to America ... to President Bush ... to Condoleezza Rice ... and to American-Lebanese field commander Satterfield ... Lebanon will not ... throw its heart to your soldiers' dogs so they will eat it ... You can make yourself heard by the commander of the American forces in the region, who is of Lebanese origin, John Abizaid ... Are you Lebanese afraid of the American naval fleets? These naval fleets have come in the past, and were defeated, and if they come again, they will be defeated again..."

Since Hezbollah's founding, its leadership has threatened America openly. In a March 1985 Newsweek article about Hezbollah, an Islamic teacher at the Bir Al-Abed Mosque in Beirut, Alia Hamden, promised a future attack by the terror organization within America. Similarly, in a July 2003 interview with the Christian Science Monitor, Shiek Nasrallah said that if America tried to dismantle his organization, American interests throughout the world will be at risk, "through any means and at any time and any place."

Al-Manar is Hezbollah's main vehicle for spreading its anti-American ideology. Such messages surface in news programs, music videos, and even game shows. For example, the question "What structure built of gray sandstone in 1792 became the source of all oppressive decisions the world over?" was asked on Al-Manar's version of "Jeopardy," "The Mission." The answer: "the White House." "The Mission" quizzes contestants from throughout the Middle East about Islamic history, geography, and arts. A spokesman for the channel, Ibrahim Musawi, explained the show's appeal "is not in an ideological way, but in an entertaining way."

On February 12, 2005, Sheikh Na'im Qassem of Hezbollah appeared on Al-Manar to discuss his hatred of President Bush: "He considers himself the god of the world. This perversion is evident in his personality ... His is patronizing and everyone abhors him ... he is a liar who tried to impose heresy on Islam." In a crowd of thousands screaming "Death to America," Sheikh Nasrallah appeared to send in May 2004 "a symbolic message that tells the Americans that we are a people that does settle for words, but is prepared for martyrdom," saying "let Bush, Powell, Rumsfeld, and all those tyrants in Washington hear ... there will only be room for great sacrifice for the call to martyrdom."

Influential Arab figures also regularly appear on Al-Manar to express anti-American sentiments. One among countless examples is the editor of the Egyptian weekly Al-Arabi, who said in an April appearance, "Anti-Americanism is like music" to his ears, calling America "the plague" and "an ongoing crime." The head of the Sunni religious courts in Lebanon, Shiek Muhammad Kan'an, called America "the garbage of all nations" in a sermon broadcast live last year.

A professor of political science at Notre Dame University in Lebanon, Dr. George Hajjar who identified himself as "coming from Columbia University" appeared on July 13, 2005, saying, "America is the New Nazism." Discussing terrorism in the West he said, "[The Americans] should be treated reciprocally. They reap what they sow." He added, "I hope that every patriotic and Islamic Arab will participate in this war, and will shift the war not only to America, but to call corners ... wherever America may be."

Anis al-Naqqash, who was involved in major terrorist attacks in Europe in the 1970s and 1980s and was released early from a French prison, appeared on Al-Manar on August 3, 2005: "The U.S. is the enemy of Arabs and Muslims ... every person must resist it ... if he can resist with weapons, it is his duty, mandated by the Koran." He also said, "Any cleric with knowledge of Islam, must ... declare Jihad against the U.S., England, and their allies."

Al-Manar's jihad is not limited to Western countries. Much of the channel's programming is devoted to anti-Semitic themes against the Jews, who are describes as "apes and pigs." Next week's column will be devoted to this subject.

Must-See Hezbollah TV: Part IV

January 25, 2006

On November 29, Al-Manar TV hosted and covered a live symposium at Lebanon's largest and only government-run university, Universite Libanaise. A student of political science, Hisham Sham'as, said, "Just like Hitler fought the Jews, we are a great Islamic nation of jihad, and we too should fight the Jews and burn them." When another student, Mahmoud Fakhri, called for Israel "to be wiped off the map," he asked the Al-Manar moderator if his statement was "too inciting." He answered, "Go ahead and incite. This is what we're looking for."

The French ban on Al-Manar for incitement against Jews in December 2004 enraged the Arab world. A spokesman for the Committee for Solidarity with Al-Manar, Ghaleb Qandil, said, "All the talk about anti-Semitism is meaningless nonsense ... we Arabs are Semites, the offspring of the 'Khazar' Jews will not be the ones to judge how Semitic we are." The Lebanese Foreign Ministry issued a statement explaining Al-Manar is "anti-Zionist, not anti-Semitic," while President Lahoud of Lebanon called the charges of "anti-Semitism" an attempt "to mislead international public opinion." As Lebanese government officials made such claims, Al-Manar aired a program about Jews spreading AIDS throughout the world.

Al-Manar's anti-Semitic outlook is derived from Hezbollah's interpretation of the Koran. The group's one-time spiritual leader, Sheikh Muhammad Hussain Fadlallah, said in 1994 that the literal text of the Koran detailing "the negative aspects of the Jews ... both in history and in the future," serves as a guide for Muslims.

A 1997 book on Hezbollah by a Reuters correspondent based in the Middle East, Hala Jaber, detailed how the terror organization's TV channel Al-Manar, regularly broadcasted excerpts from the groups original manifesto. Citing Surat al-Ma'idah, Verse 82 of the Koran, the manifesto said Hezbollah's "jihad" is a "religious obligation" and that "the animosity between Muslims and Jews goes back to the early days of Islam."

Al-Manar children's programming also includes anti-Semitic themes taken from Islamic teachings. For example, a claymation special from December 7 was titled "Stories from the Koran." The program was based on a famous Hadith from Islamic history in which the Jews became apes and pigs.

The head of Radio Islam in Sweden and one of Europe's most notorious anti-Semitic Muslims, Ahmad Rami, appeared on Al-Manar on September 30 to explain that his beliefs are based on what "the Koran says, that our battle is with the Jews" and that "Judaism is a criminal and dangerous mafia."

Anti-Semitic incitement on Al-Manar is not drawn solely from interpretations of the Koran, but from classic anti-Semitic subject matter including the blood libel; Protocols of the Elders of Zion; Nazi forgeries, and attacks on Judaism.

Explaining how the Jews have "distorted the Torah," the Mufti of Tripoli, Sheikh Taha Al-Sabonji appeared on Al-Manar on April 22, 2004, saying: "Those responsible for all civil strife ... throughout history were the Jews. This is verified by anyone who has read Jewish literature and ... the Koran." On the same show, the secretary-general of the Islamic Universities Association, Dr. Ja'far Abd Al-Salim responded: "The Protocols of the Learned Elders of Zion clearly refer to this."

The president of the Middle East Center for Studies and Public Relations, Hisham Jaber appeared on Al-Manar on July 11 to discuss the vileness of the Jews claiming they were behind the attacks of September 11, 2001. He also said, "We know that since the Protocols of the Elders of Zion, Zionism has forged the New Testament. ... 60 million in the U.S. alone have left Christianity to become believers in the Torah. Global Zionism has tried to forge the Holy Koran."

A Lebanese journalist, Arafat Nizam Al-Din, quoted Nazi propaganda during a November 11, 2004, interviewed on Al-Manar. He added his on twist on American history and Jews with fictional anti-Semitic quotes from President Washington, President Jefferson, and Benjamin Franklin. Similarly, Al-Manar aired a discussion by Dr. Ghazi Rabab'a, Professor of Political Science at Jordan University, on May 27, 2004: "President Jefferson told the American people in an official speech, 'If you do not expel the Jews from your land, they will enslave you. ... If we open the Talmud, the Jews' false book, we see that."

Another one of the countless examples of such anti-Semitism on Al-Manar (which can be viewed at www.memritv.org) comes from the deputy head of the Palestinian Clerics Association, Sheikh Muhammad Ali, who appeared on August 19 and spoke of wiping out the Jews, how the Zionists have forged the Torah and Talmud, and quoted directly from the Koran to justify killing Jews.

Next week's column will explore the Iranian connection to Al-Manar.

Must-See Hezbollah TV: Part V

February 1, 2006

"To deny the Iranian aid issued to Hezbollah would be like denying that the sun provides light to the earth."

— HEZBOLLAH'S ORIGINAL SECRETARY-GENERAL, SHEIKH SUBHI TUFEILI

On February 16, 1985, Hezbollah released its manifesto bearing a picture of "our leader," the Ayatollah Khomeini on the back cover. A Washington Post article from that year reported on, "Iranian Revolutionary Guards carrying out their missionary work, indoctrinating the Lebanese Shiites in the spiritual and political teachings of Khomeini."

Iran's ideological and financial support of Hezbollah was detailed in Hala Jaber's 1997 book on the terror organization: "Hezbollah is coy about revealing the sums it has received from Iran." She added, "reports have spoken on figures ranging from $5-$10 million per month."

Hezbollah's most important tool in spreading its ideology of death and celebration of martyrdom - heavily influenced by the Iranian theocracy - is the TV channel Al-Manar. With a multimillion-dollar headquarters in Beirut and as the Christian Science Monitor has reported, a budget that grew tenfold between 1991 and 2001 (some estimate its current annual budget to be between $10-$50 million), that the Iranian government is backing Al-Manar should come as no surprise. According to an article published by in the Transnational Broadcasting Studies' winter-fall 2002 issue, "Iranian ayatollahs backed and helped to launch Al-Manar" and the channel's first broadcast was of the 1989 funeral of the Ayatollah Khomeini.

In a June 2, 2002, interview on Al-Manar with the deputy secretary-general of Hezbollah, Sheikh Na'im Qasem, the sheikh quoted the Iranian regime in explaining his organization's position on Israel and jihad: "Muslims should annihilate Israel. ... Imam Khomeini said, 'The goal of this virus [Israel] that was planted in the heart of the Islamic world. ... The danger is to the whole Middle East... and the solution is in annihilating the virus.'" Mr. Qasem also described how good it was to see "a mother saying goodbye to her son, awaiting his return as a shahid [suicide bomber]"

Hezbollah's leader, Sheikh Hassan Nassrallah, gave a speech on February 19, broadcast live on Al-Manar containing both a threat to America and a salute to Iranian Supreme Leader Khamenei: "If America ... stops its aggression ... we will have no problem with it. We don't want to go to Washington to fight America." He added, "How can death become joyous ... sweeter than honey? Only through conviction, ideology, and faith ... as the Leader Imam Khamenei said ... the most honorable killing and the most glorious martyrdom is when a man is killed for the sake of Allah, by the enemies of Allah, the murderers of the prophets [i.e. the Jews]."

Ms. Jaber's book also detailed how Hezbollah TV instructs Muslim youth to terrorism: "Al-Manar ... is dominated by religious programs. Pictures and names of martyrs are screened, supported by verses from the Koran which glorify such deaths. The aim is simple, to indoctrinate the minds of the young ... with the idea that those seek martyrdom will be rewarded with more pleasure than can ever be achieved during this earthly lifetime."

Al-Arabiya TV aired a program on the celebration of martyrdom by Hezbollah and Al-Manar on August 19. In one scene, a young boy is shown viewing footage of a suicide bomber in a car that exploded. The boy said: "I love to watch him," explaining it was his father.

The mother of martyr Bassel Al-Din appeared on Al-Manar on May 22. She cried in happiness when telling the channel: "Bassel had a wish. ... Whenever I told him I wanted to marry him off, he would say, 'Yes, mother, you'll marry me off like this in paradise.' And indeed, the martyr Bassel got married in paradise. I congratulate the black-eyed virgins who took Bassel from me."

A November 11, 2004, "Mother's Day Special" on Al-Manar featured comments from many mothers of martyrs. One stated, "All I want is martyrdom. I'm willing for all my children to become martyrs." Another said, "It's true I sacrificed a son, but others have sacrificed two or three. I hope more of my sons will become martyrs." The Al-Manar moderator praised them and explained, "The reward of ... all martyrs' mothers is not in vain. ... Not only locally, this is an experience that is now shared by all societies."

Just this past week, Dutch authorities blocked the transmission of Al-Manar for spreading hate and stated the channel encourages the radicalization of Muslims and glorifies terrorist attacks. Next week's column, the conclusion of this series on Al-Manar, will explore Western reaction to Hezbollah's attempt to incite Muslims in the West via its TV channel.

Must-See Hezbollah TV: Part VI

February 9, 2006

Since Hezbollah's TV channel Al-Manar began broadcasting via satellite in 2000, it has been at the center of controversy throughout the West. In America, Canada, France, Australia, Spain, the Netherlands, and elsewhere, the channel has been banned.

In 2003, Australia was first to censure Al-Manar. The Australian daily newspaper said by way of explanation, "the channel incites to terrorism." During a Swedish parliament session on March 18, 2004, Mikael Oscarsson of the Christian Democratic Party asked Prime Minister Goran Persson to put an end to Al-Manar broadcasts in his country, describing them as "appalling propaganda of incitement" that "can only be compared with that of the Nazis."

The French government banned Al-Manar in December 2004 for violating repeatedly the country's anti-hate laws. America and Canada followed just days afterward. A State Department spokesman, Richard Boucher, said, "We don't see why ... a terrorist organization should be allowed to spread its hatred and incitement through the television airwaves."

The Spanish government blocked transmission of Al-Manar on July 14, 2005. A conservative foreign affairs spokesman at the European Union, Charles Tannock, responded as follows: "Hezbollah uses Al-Manar to spread hatred and incite people to commit terrorist acts against innocent civilians ... every effort must be now made by the E.U. to prevent further brainwashing of vulnerable young people by fundamentalist religious extremists." He said Hezbollah ought "not be allowed to spread hate-filled propaganda on our continent."

Most recently on January 26, Dutch authorities blocked the transmission of Al-Manar for spreading hate, saying the channel encourages the radicalization of Muslims and glorifies terrorist attacks. Justice Minister Piet Hein Donner was quoted by the AFP as calling "for a European Union-wide solution to the problem of such television channels."

Some Arabs have also been critical of Al-Manar. In an interview with Al-Jazeera on July 26, 2004, a former Iraqi government minister, Hoshyan Zebari, accused

Arab satellite channels of inciting violence - singling out Al-Manar. In November, the editor of the London Arabic daily Al-Sharq Al-Awsat, Tariq Al-Homayed, criticized "ideology-laden satellite channels"for causing damage to the Arab world, mentioning "Hezbollah's channel Al-Manar."

The channel does receive tremendous support from the Arab world and in particular the Lebanese government. After the American ban, a former Lebanese premier, Salim Hoss, said, "the U.S.move was akin to granting Al-Manar a medal of honor. It is an official recognition of its genuine effectiveness."

Lebanese government officials have explicitly stated Al-Manar is formally under its control. President Lahoud said on July 1, "Al-Manar is a national Lebanese consensus issue, and it is protected by Lebanese laws." The director general of the Information Ministry, Hassan Falha, in August 2004 said, "Al-Manar is a Lebanese media institution working in abidance with the Lebanese laws and laws regulating media in Lebanon. The Lebanese government is concerned about the Al-Manar issues at all press, legal, diplomatic, and political levels."

Following Al-Manar's ban in the West, the editor-in-chief of the channel's news division, Abdullah Shamseddine, told the Daily Star on December 20, 2004, "The damage is merely political ...Viewers can still watch via NileSat and by adjusting their dishes to receive our signal. The same applies to South American countries."

This strategy has not gone unnoticed by Western governments. As an E.U. spokesman, Charles Tannock, said on July 14, 2005, "I have asked the British Presidency of the E.U. to raise the matter with Egypt and Saudi Arabia, both of whom continue to operate satellites which broadcast Al-Manar across Europe."

Over the past six weeks, this column has detailed Al-Manar's history; its anti-American stance; its connection to Hezbollah, and its use by Hezbollah to incite Muslims throughout the world. Just days ago, Hezbollah leader Sheikh Nasrallah made a statement covered by Al-Manar about the Muhammad caricature controversy, which could be interpreted as being directed toward Muslims in the West.

"If there had been a Muslim to carry out Imam Khomeini's fatwa against the renegade Salman Rushdie, this rabble who insult our Prophet Muhammad in Denmark, Norway, and France would not have dared to do so ... I am sure there are millions of Muslims who are ready to give their lives to defend our prophet's honor and we have to be ready to do anything for that," he said.

This quote represents the danger posed by Hezbollah and its TV channel. One solution is for the Lebanese government to pull the plug on Al-Manar.

Iran Attempts To Stifle Internal Dissent

February 9, 2006

Iranian opposition to President Ahmadinejad is on the rise following a series of recent setbacks, including election losses for his candidates and criticism of his handling of the nuclear crisis. As discontent increases, however, Mr. Ahmadinejad's regime is going to spectacular lengths to purge and silence its opposition, much as Nazi Germany did in consolidating its power.

This purge includes the banning of books, reformist papers, Web logs, and satellite TV channels, as well as working against groups such as labor unions and student and women's organizations.

In the first week of December, Iran cut off its citizens' access to YouTube and Amazon.com; it had earlier blocked Wikipedia and the New York Times. The Iranian authorities did this, they said, to "purge the country of Western cultural influences."

In the reformist daily Rooz of January 25, Nooshabeh Amiri discussed the attacks on freedom in Iran and warned that the country will "become a sanctuary for devils and beasts."

"Iran is the battlefield of evil forces; as if all of the devils that came into existence since the initial days of the revolution have now met at the same place in this historical juncture," he wrote. The Iranian regime can "instill terror in society," he added.

Since Mr. Ahmadinejad came to power, a number of student groups that oppose his policies have spoken out against him and have been attacked in response. During a speech by the president at Amir Kabir University on December 11, students burned his picture and shouted, "Death to the dictator."

One student leader who was blocked from attending his university, Abbas Hakimzadeh, spoke to Rooz on December 4 about government activity against the student movement. The regime is banning student publications, bugging student leaders, maintaining surveillance on them, threatening their families, and even expelling and arresting some of them, he said. But Mr. Hakimzadeh said he was optimistic that "the hardliners" in Iran will lose in their attacks against student organizations, which "act like a watchdog over civil institutions."

"Under the current conditions, when every section of society suffers from hopelessness, having faith and hope is critical, as this leads to action. Look at how during the last year the government has driven, like a bulldozer, to crush civil organizations, and ignores the wishes of people," Mr. Hakimzadeh said. Still, "those forces that have not lost hope continue to hold this flame of action and faith high."

Dissent among university faculty has also been the focus of the Iranian regime's wrath. According to a report on the Advar News Web site last month, a new "trend" has been to force any faculty member who speaks out against the regime into retirement.

Women's rights activists in Iran have been another target of the government. At the end of January, four women were arrested on their way to a journalism workshop. The father of one of the women, Farnaz Seyfi, reported on the details of the arrest on his Web log, Emshaspenda.

"They were interrogated on issues related to the campaign known as '1 million signatures demanding changes to discriminatory laws,'" one of the lawyers representing the four told the Iranian Students' New Agency. "The campaign was initiated earlier this year as a grassroots effort to educate the public and raise awareness about gender discrimination in Iran."

Other women's rights activists have been harassed this year and "accused of undermining national security and cooperating with the enemy." Male journalists — including intellectual Ramin Jahanbegloo and Ali Farahbakash, who reportedly is sick in prison — have been arrested on the same charge.

Omid Memarian provided a detailed account of how the Iranian government extracts confessions from its opponents "through torture" and "intimidation" in Rooz on November 29. In one incident, in which he was taken into custody, he wrote: "The judiciary tsar, Ayatollah Shahrudi, showed us images from the Internet where the heads of some the country's religious leaders had been superimposed on bodies of other people in compromising situations. We were appalled and I made it clear that we had not done that."

"That incident demonstrated how easy it is for the intelligence personnel to mislead senior decision-makers of the country simply because of their own tunnel vision and dogmatic views of the world," Mr. Memarian wrote. "This example may also demonstrate how and through what mechanisms Ayatollah Khamenei is convinced that 'foreign hands' are threatening the traditional culture of the land, or that the Internet is an 'enemy.'"

As Iran takes another page from Nazi Germany, one must ask where the Western student groups, women's right activists, union members, journalists, and other freedom-seeking individuals are, when their Iranian counterparts desperately need support. Many things are possible through the Internet and e-mail.

The Arab Press on Evangelical Christians

February 15, 2006

Attacks on Evangelical Christians by government-controlled Arab press outlets have received too little attention by the West. In a recent example in a leading Egyptian weekly, an influential professor wrote that Evangelicals in America could pose a greater danger to the world than Nazi Germany.

Arab television also frequently berates Evangelicals. Programs on Saudi Iqra TV and Hezbollah's Al-Manar TV have featured guests saying Evangelicals were involved in planning the attacks of September 11, 2001. The most popular form of defamation of Evangelicals, however, centers on President Bush's religious beliefs.

In an article titled "Impeaching Bush" in the January 5-11 edition of an Egyptian government weekly, Al-Ahram, a political science professor at Cairo University, Hassan Nafaa, expressed his displeasure with the results of the 2004 election. He said if "the American people [would] decide to oust the current president ... this decision [would have heralded] the end of the age of the neo-cons and perhaps the beginning of the collapse of their Evangelist Imperialist project."

The article also addressed Mr. Bush's becoming a born-again Christian: "Suddenly he was gripped with the desire to repent and to be born again ... a group of neo-con radicals latched onto him and began to mould him to serve their power-hungry designs which entailed capitalizing on the growing tide of American Christian fundamentalism..." Ms. Nafaa ended his column by calling for Mr. Bush, who is "more dangerous than Osama bin Laden" to be removed from power.

In an Al-Ahram article from the weekly's October 6-12, 2005 edition, Mr. Nafaa spoke to "the decline of liberalism," in the West, which he connected to American Evangelicals: "The collapse of the USSR opened the path for the Right in America, particularly its extreme Christian fundamentalist wing. This wing has local and international ambitions at the core of which is a discriminatory and racist agenda that poses a greater danger to political liberalism than communism and, perhaps, even Nazism."

An article titled "The Real Axis of Evil" by A. Al-Agha, which appeared on December 18 in Syria's leading government daily, Teshreen, the writer mentioned an "axis of evil" inside America: "Fundamental Christianity, or if you prefer the Evangelicals in the United States, can be termed one part of this axis ... It would not be a stretch to claim that religion in the United States and most likely throughout the areas of the world where the United States holds sway is Zionist controlled ... Anything Islamic or of Arab origin is deemed evil ."

Influential Arab and Muslim Americans such as a professor at the University of Delaware and Fellow at the influential Brookings Institution's Saban Center, Muqtedar Khan, often write about American Evangelicals in the Arab press. In the June 5-11, 2003 edition of Al-Ahram Mr. Khan said, "One of the reasons for the growing Islamophobia in this country is the anti-Muslim rhetoric coming from the Evangelical Christian community." He added, "I must remind readers that hate mongering is not common in the Christian communities of North America. It is indeed a rare but egregious blemish found only among the Evangelists."

In an article published in the American Muslim Minaret Magazine in July 2003, Mr. Khan suggested Evangelicals' problem with Muslims lies in mass conversions of Americans to Islam: "Christian Evangelists are seeking to stem the tide of conversions to Islam. By demonizing Muhammad as a pedophile and a polygamist they hope to generate enough distaste for Islam that it becomes difficult for more open and tolerant Americans to be accepting of Muslims and Islam. They also hope that they can reverse the growth of Islam in America ... attracting thousands of converts ... While the traffic from Christianity to Islam is very heavy, the reverse flow, much to the frustration of the Evangelists, is barely a trickle. This tirade of hate from the Evangelists is merely the continuation of a very old Christian preoccupation with the demonization of Muhammad. Christian thinkers have always abused the Prophet of Islam. A famous example is Dante, who in his 'Inferno' ... throws him in the bowels of hell."

The Arab government-controlled press's contempt for non-Muslims is not limited to Evangelicals. Hatred of other Christians, Jews, Hindus, and others can also be found daily.

Al-Jazeera Provokes Cartoon Rage

February 22, 2006

The story behind how cartoons of Muhammad have inflamed the Muslim world is well-known. What is less well-known is how Al-Jazeera added fuel to the fire. As a February 8 article in a London Arabic daily, Al-Sharq Al-Awsat, reported, "When the controversy began to gain momentum, the Qatar based Al-Jazeera increased its [distorted] coverage."

The article said, "the issue disappeared from the radar until Sheikh Youssef Al-Qaradhawi, the mufti of Al-Jazeera, seized upon it and called for Muslims worldwide to protest." Sheikh Al-Qaradhawi, who is a spiritual leader to millions of Muslims worldwide, told his followers how to respond to the cartoon controversy in his February 3 sermon: "The nation must rage in anger. It is told that Imam Al-Shafi said, 'Whoever was angered and did not rage is a jackass.' ... We are not jackasses for riding, but lions that roar ... We are a nation that should rage for the sake of Allah ... and show our rage to the world."

The Muslim nation did indeed "rage," and it was covered by Al-Jazeera. A demonstration opposite a Kentucky Fried Chicken in Damascus, Syria, which aired on January 28, showed a crowd chanting, "Oh, Arab, wherever you are - destroy the army of the Americans ... Beware, oh America ... We've hammered the first nail in the coffin of colonialism." On February 4, Al-Jazeera filmed rioters actually lighting the Norwegian embassy in Damascus on fire (to view, visit www.memritv.org).

The network also televised influential Arabs making threats against the West. For example, on January 30, a group of Kuwaiti members of parliament appeared to discuss the cartoons. One said, "When [Prophet Muhammad] is attacked by the dogs of the extreme right, who suckle from the breasts of Zionism, this causes him no harm ... These European perverts and their governments ... these barking dogs and filthy pigs do not surprise us. I use these terms because they suit them. These terms befit their vileness, despicability, filth, and impurity. You know this, because you have traveled to Europe." Another Kuwaiti member of parliament said, "We should not be surprised if some journalist or newspaper distorts the image of Islam. But what does surprise us is that the [Danish] prime minister says this is freedom [of expression]. If freedom there reaches such a degree, they should consider the bombings that take place to be freedom as well."

On February 4, Al-Jazeera covered a speech meant to incite fellow Muslims about the cartoons by the chairman of the Jordanian Lawyers Association, Saleh Al-'Armuti: "My beloved Prophet of Allah, I will sacrifice my soul for you. I will sacrifice my blood and my family for you. ... Where are you, oh Muslims? Where are you, oh Arabs?"

Hamas leader Khaled Mash'al's address at the Al-Murabit Mosque in Damascus aired live on February 3 with a message to the West: "We apologize to our Prophet Muhammad, but we say to him: Oh Prophet of Allah, do not be saddened, your nation will be victorious. We say to this West, which does not ... learn its lessons: By Allah, you will be defeated ... They do not understand the Arab or Muslim mentality, which rejects the foreigner ... I say to the [European countries]: Hurry up and apologize ... because if you do not, you will regret it ... Tomorrow our nation will sit on the throne of the world ... Today you are fighting the army of Allah. You are fighting against peoples, for whom death for the sake of Allah ... is preferable to life. You are fighting a nation that doesn't not tire, even after a 1,000 years of fighting ... You will be the ones to be defeated."

Hezbollah leader Hassan Nasrallah's response to the cartoons was covered by the Arab press, including Al-Jazeera on February 2 and 3. He said, "If any Muslim had carried out the fatwa of Imam Khomeini against the apostate Salman Rushdie, those despicable people would not have dared to insult the Prophet Muhammad - not in Denmark, Norway, and France ... This is about the Prophet of ... 1.4 billion Muslims. I am certain that ... hundreds of millions of Muslims are ... willing to sacrifice their lives in order to defend the honor of their Prophet."

It is well-known that captured terrorists in Iraq have admitted watching Al-Jazeera inspired them to perform jihad. It could be argued that the network's coverage of the cartoon controversy has also incited viewers throughout the Muslim world. As worldwide protests continue, it is now important to scrutinize Al-Jazeera for what it is airing.

Muhammad Cartoon And Samarra Shrine

March 1, 2006

Within hours of this week's attack on the Shiite holy shrine in Samarra, Iran's supreme leader, Ayatollah Ali Khamenei, blamed it on "intelligence bodies of Iraq's occupiers and the Zionists." Hours later Hezbollah's leader, Sheikh Hassan Nasrallah, appeared on Al-Manar TV also blaming the attack on "U.S. occupation authorities," while President Ahmadinejad of Iran was certain it was the work of the "defeated Zionists and [American] occupiers."

During a press conference with the Syrian prime minister, Iran's vice president, Parviz Davoudi, claimed that the "Zionist hands and the occupiers' conspiracy are visible in this crime." On February 26, Iran's Foreign Ministry spokesman, Hamid-Reza Asefi, also stated: "We see the hand of the U.S. Ambassador and the Zionist regime in the blast."

A leading Islamist, Sheikh Al-Qaradawi, pointed to "the U.S. occupation and the lurking Zionist enemy," while a radical Iraqi leader, Muqtada Al-Sadr, accused the American or Iraqi government of blowing up the shrine.

Within hours of the attack, the Tehran Times political desk issued a report that read, "the occupying U.S. regime ... is the main element responsible for these criminal acts," while a leading Shiite Pakistani leader who is the president of Millat-e Jafria Pakistani, Allam Ali Naqui, issued a statement to IRNA explaining the bombing was "a ploy to divert attention of the Muslim world from the Imperialist forces' diabolical designs against Islam."

The immediacy surrounding the emergence of conspiracies about the bombing of the Samarra shrine as well as what occurred after world-wide protests regarding the publication of the Muhammad cartoons should come as no surprise. Influential Arab and Iranian figures were quick to blame the usual suspects: Jews; Christians; Zionists, and America.

A Qatar University lecturer, Ali Muhi Al-Din Al-Qardaghi, appeared on Al-Jazeera on February 3 to discuss a Zionist and Christian conspiracy: "We must realize that something bigger than this [cartoon] incident lies behind this campaign. Zionism and the extreme Christian right are behind it ... This is a Crusader Zionist campaign, which is led by the extremist pro-Zionist right, headed by George Bush in America ... This campaign is an attempt to arouse religious, Crusader fanaticism among Europeans."

The new Palestinian Authority ambassador in Washington, D.C., Afif Safieh, appeared on CNN's "Late Edition With Wolf Blitzer" on February 12. He blamed the cartoon controversy on "the pro-Israeli Likud wing around the world who wants to ... put on a collision course the Western mainly Christian world."

In a Friday sermon by Ayatollah Khamenei that aired on Iran's Channel 1 on February 7, he charged: "That a profound Zionist plot is at the core of the matter ... This Crusade is the war of the Christian peoples against the Muslim peoples ... Today, I see the Zionist hands are preparing the ground in the Christian countries for this ... They also control the newspapers ... They are the ones who completely control the American government."

According to a report in Iran's Mehr News Agency on February 19, Ayatollah Ardebili explained the printing of the cartoons was "part of a Zionist plot to start a new crusade," while Ayatollah Nuri Hamedani also said it was "international Zionism's attempt to sow dissension between Christianity and Islam."

The secretary of Iran's Expediency Council, Moshen Reza'i, discussed the cartoon controversy, calling it an "American-Israeli Project," on February 7 in an appearance on Iranian TV 2, while an article in the Tehran Times of February 4 added: "A careful analysis of similar events around the world over the past five years reveals that the U.S. neoconservatives and the Zionist lobby have formulated a plot to influence public opinion in the West so as to foster animosity between Islam and Christianity ... The propaganda campaign inspired certain Christian fundamentalist groups to launch an ideological offensive against Islam."

IRNA reported on February 13 that a group of Iranians protesting outside the French Embassy in Tehran declared, "We believe the move of the European press in insulting the holy prophet of Islam is a gesture designed by Zionist lobbies and supported by governments of the countries in line with the West's warmongering polices." The crowd of Iranian protesters then chanted: "Nuclear energy is our inalienable right, Death to the U.S., Death to France, Death to Israel!"

No matter what tragedy - whether natural or man-made - next plays out in the Middle East, it can be expected that influential Arabs and Iranians will come up with conspiracy theories to explain them. Unfortunately, few voices there have denounced those who are blaming the Jews and Christians for every misfortune that has befallen the Muslim world.

Staging Israel's Destruction

March 8, 2006

Since its founding, the terrorist organization Hamas has stated openly that its ultimate goal - the destruction of Israel - will unfold in stages. As the Hamas charter, written on August 18, 1988, reads, "The Islamic Resistance Movement is well qualified for the upcoming stage [of the struggle] with the Jews."

During the Palestinian-Arab elections, questions arose regarding whether Hamas's plans for the step-by-step destruction of Israel had changed since it the group's political platform did not explicitly call for such ruin. A Hamas spokesman, Sami Abu Zuhri, denied the group's purpose had been modified: "The platform refers to details and implementation methods for the next four years, while the charter lays out our permanent strategic views."

One Hamas candidate, Salah Al-Bardawil, told Islamonline on January 14, "The platform presents a realistic view that reflects Hamas's goals for the next four years. Had we spoken of eliminating and eradicating Israel within this period, we would have been deceiving our people ... But this does not stand in contradiction [of the fact that] we place emphasis on the elimination ... of Israel."

In a February 13 interview with Dream TV, the deputy head of Hamas, Musa Abu Marzouq, was asked if "Hamas is talking in two voices - to the Palestinians it says 1948, and to the foreigners it says 1967." He answered, "From the [Jordan] River to the [Mediterranean] Sea belongs to the Palestinians ..."and "we also say that we are dealing realistically with the current phase - an independent Palestinian state with full sovereignty over the West Bank, Jerusalem, and Gaza." Mr. Marzouq was then asked: "Is this a partial or phased solution?" to which he replied, "It is a temporary and phased solution. This is not the permanent solution."

At a January 29 press conference covered by Al-Jazeera, Hamas's leader, Khaled Mash'al, also discussed Hamas's philosophy. "We are realistic, and know things are done gradually, in stages," he said. He explained that Hamas understands that its "strategic goals" to "abolish Israel" cannot be accomplished in "a matter of moments." But added: "We believe in acting according to stages, gradually and realistically ... step by step. We will establish our state on any piece of land we liberate."

A Hamas representative in Lebanon, Osama Hamdan, gave an interview to ANB TV on January 26 in which he explicitly discussed the "phased solution" to destroy Israel by liberating "Acre, the Galilee, Haifa, Jaffa, Lod, Ramla, and Beersheba ... What [Hamas leader] Isma'il Haniya said about accepting a phased solution - or what we called 'liberation in phases' is possible."

In an August 18 interview with Asharq Al-Awsat, Hamas's leader Dr. Mahmoud Al-Zahar elaborated on Hamas's strategy, saying, "Our plan in the first stage is to liberate the lands occupied in 1967." In an interview broadcast by Al-Manar TV on January 25, Mr. Al-Zahar said, "Palestine means Palestine in its entirety - from the [Mediterranean] Sea to the [Jordan] River ... If at present we cannot bring about a decisive victory ... we are saying that we may agree to found a state on any part of the land."

During the week of Israel's withdrawal from Gaza, Hamas Friday sermons considered the move to be a stage leading toward Israel's dissolution. Sheikh Mazzar Rayan on September 16 said, "The vanquishing of the enemy in Gaza does not mean this stage has ended. We still have Jerusalem and the pure West Bank ... We do not distinguish between what was occupied in the 1940s and what was occupied in the 1960s. Our jihad continues."

A Hamas spokesman, Mushir Al-Masri, explained in an interview on Al-Arabiya TV on December 5, 2004, "I think that the strategy of Hamas is clear: liberating the entire land of Palestine from the plunderers. Hamas ... believes that Tel Aviv is like Gaza."

Speaking to a crowd chanting, "Destroy Tel Aviv," broadcast on Al-Jazeera on December 20, Hamas leader Khaled Mash'al told the crowd: "When will we celebrate ... the liberation of the West Bank? ... the liberation of Jaffa, Haifa, Safed, and the Negev? When will we celebrate the departure of the last Zionist from our land?... It could be soon."

Hamas is not alone in its planned "staged" destruction of Israel. Following its election victory, Mr. Mash'al received a phone call from Iranian President Ahmadinejad, who was quoted by IRNA on January 30 as saying: "We thank God that struggles, self-sacrifices and martyrdom-seeking attempts of the children of [the] Palestinian nation have today resulted in the first stage of victory."

The next stage is well-known. It is Mr. Ahmadinejad's call for Israel to be wiped off the map - a dream shared by Hamas.

Hamas's Philosophy on Raising Martyrs

March 15, 2006

A mother of five martyrs, Umm Nidal, says that when a child blows him or herself up, "he is not throwing himself to death or to perdition ... it is called martyrdom." Ms. Nidal has become a darling of the Arab press and a member of Palestinian Authority Legislative Council - during public appearances, she calls on other Muslim mothers to "sacrifice" their children for "jihad." In a December 21 interview aired by Dream TV, Ms. Nidal said the Koran supports her children's quest for martyrdom, and said she would gladly "sacrifice" her five remaining sons.

Ms. Nidal's views are shared by many who voted her and her party, Hamas, into the Palestinian government, and should be taken seriously.

Tomorrow on Capitol Hill, the Middle East Media Research Institute will hold a briefing titled, "Hamas: Its Ideology and Strategy For the Future," sponsored by the House of Representatives International Relations Subcommittee on International Terrorism and Nonproliferation.

Translated remarks by Hamas leaders broadcasted on Arab TV, will be viewed and the terror organization's publications including its covenant will be examined.

Hamas's strategy to advance its culture of violence, which relies heavily on imposing its ideology on future generations through education and child-rearing, will also be examined. The following translations are examples of how Hamas leaders express the importance of including women and children in jihad:

* "Jihad becomes a personal duty of every Muslim ... This requires the propagation of Islamic consciousness among the masses ... The indoctrination campaign must involve ulama, educators, teachers ... It is [also] necessary to introduce essential changes in the curricula, in order to eliminate the influences of the intellectual invasion which were inflicted upon them by the Orientalists and the missionaries."

- Article 15 of the Hamas covenant

* "It is necessary to educate the next Islamic generations in our region in an Islamic way ..." and all Muslims must participate in Jihad.

  - Article 16 of the Hamas covenant

* "Our education system will not represent Palestine as a coastal strip stretching from Rafah to Beit Hanoun ... Our ministry of culture will teach them how the martyr is turned into prose, literature, and poetry ... Palestine means Palestine in its entirety - from the [Mediterranean] Sea to the [Jordan] River..."

  - Hamas leader Mahmoud Al-Zahar; from comments broadcast January 25 on Al-Manar

* "Hamas has tens of martyrs who are willing to carry out attacks ... An operation of such martyrs exceeds that of Arab armies ... The importance of the weapon of such martyrs is no less than the importance of nuclear weapons."

  - Hamas leader Khaled Mash'al; extracted from his speech to a Palestinian Arab high school students in June 2001

* "I thank Allah ... for His support in the Jihad ... and I ask him to help us to liberate Jerusalem ... Acre, Haifa, Jaffa, Safed, Nazareth, Ashkelon, and all of Palestine ... We remember the martyrs ... Who will forget ... the children that you sacrificed? ... We saw our expectations [realized] ... until [even] women urged their sons to commit martyrdom ... and the fathers received the message of the sons' martyrdom in praise and cries of 'Allah Akbar."

  - Hamas military commander Muhammd Deif; from an audio cassette released August 27

* Palestinian Arab parents who encourage their children to become suicide bombers are "people [who] compete with one another for martyrdom, people who encourage their sons to die ... A mother readies her son, and then receives the news of his martyrdom with cries of joy. A father receives the news of his son's martyrdom with pride and honor. Our women push our children [towards martyrdom]..."

  - Extracted from a September 16 Hamas sermon covered by Al-Jazeera

* "The role of the Muslim woman in the war of liberation is no less important than that of the man, for she is the maker of men. Her role in guiding and educating the next generation is very important..."

  - Article 17 of the Hamas covenant, on women's role in preparing children for Jihad

The event will also look at the prospects for peace between Israel and the Palestinian Arabs. As Umm Nidal told Dream TV on December 21: "Peace means the liberation of all of Palestine, from the [Jordan] to the [Mediterranean] Sea. When this is accomplished - if they want peace, we will be ready."

The Season's Top 10 Conspiracies in the Arab and Iranian Press

March 22, 2006

As this column has documented extensively, conspiracy theories dreamed up by Arabs and Iranians appear daily in their press. Recent events highlighting the extent of this phenomenon include their commentary on the destruction of the Shiite shrine in Samarra and the controversy surrounding cartoons of Islam's prophet Muhammad.

The Samarra attack was blamed on America and Israel by - among others - Iran's supreme leader, Ayatollah Sayyid Ali Khamenei; Hezbollah's leader, Hassan Nasrallah; Iran's president, Mahmoud Ahmadinejad, and a leading Islamist, Sheikh Al-Qaradawi.

On February 27, an editorial in the Iranian daily Jomhuri-ye Eslami laid out in detail why America blew up the shrine, while a February 28 article in the Tehran Times titled "Who Was Behind the Samarra Blasts?" held responsible "the unholy alliance of the U.S., Zionists, and the terrorists operating in Iraq."

It was no surprise on whom the cartoon controversy was blamed. A vocal Swedish anti-Semite, Ahmad Rami, pointed to "the Jewish-controlled press" in the Tehran Times on February 23, while in Lebanon, the vice president of the Higher Shiite Council, Sheikh Abdel-Amir Qabalan, was quoted by the Daily Star on February 11 as saying "the Zionists" initiated the uproar.

The following are the top 10 conspiracies from the Arab and Iranian press this season:

10) "Israeli security organizations are in efforts to spread flagrancy and corruption among young [Iranians] to harm the Islamic establishment" and Israel has been behind the smuggling of liquor into Iran, the Islamic Republic News Agency quoted an Iranian police commander, Colonel Hossein Abdi, as saying on January 31. A Syrian government daily, Teshreen, on January 8 quoted the Iranian foreign minister condemning an unnamed Zionist company "that published certain cards insulting Muslims and their religious traditions in Iran."

9) "The U.S. propagates a long-term plan for conflict with Muslims. The plan designates no countries, cultures, [or] history ... It is war against some unidentifiable mass entity, where the only visible landmarks are vital ports and strategic natural resources," the Syrian minister of expatriates, Bouthaina Shabban, wrote in a February 22 column in the London Arabic daily Al-Sharq Al-Awsat.

8) America instructed some Muslim countries to "eliminate" verses from the Koran, Ayatollah Nuri Hamadani told elite members of the Iranian military, according to a Tehran Iranian Labor News report released the first week of January.

7) An article in the Egyptian government Al-Ahram Weekly of March 2-8 by Khaled Amayrch reported on Israeli religious political party platforms for dealing with the Palestinian Arabs: "enslavement whereby non-Jews living under Jewish law are forced to become 'water carries and wood hewers,' expulsion, or outright extermination."

6) Turkish Muslims discovered America, according to documents presented by a Turkish professor, Ruat Sezgin, who founded and chairs the Institute for the History of Arabic-Islamic Science, at the Association of Written Heritage office in Tehran on February 5.

5) "British Hand Behind Ahvaz bombings," a headline published by the Mehr News Agency on January 26 read. Iran's foreign minister expressed the same sentiment when speaking to Al-Jazeera on January 25; in a January 26 interview with the network, Iran's interior minister claimed America, Britain, and Israel were behind a series of plane crashes in Iran.

4) The Mossad has planned terror attacks in Lebanon, the January 31 edition of Teshreen quoted President Lahoud of Lebanon as saying, while a Qatari daily, Al-Raia, on January 26 quoted a former Syrian defense minister, Mustafa Tlas, as saying Israel assassinated a former Lebanese prime minister, Rafik Hariri.

3) Small teams of CIA-sponsored militias are behind all the kidnapping and killing of foreigners in Iraq, Firas Al-Atraqchi wrote in an extensive report in the Al-Ahram Weekly of February 9-15. It included a picture of a kidnapped Christian Science Monitor reporter, Jill Carroll.

2) "The Jewish Walt Disney Company" created the Tom and Jerry cartoon to improve the image of Jews, because Jews were called "Dirty Mice" in Europe,

an Iranian government adviser, Hasan Bolkhari, said February 19 on Iranian TV Channel 4.

1) Israel created the avian flu virus in order to damage "genes carried only by Arabs," a January 31 column by Abd Al-Rahman Ghunwym in a Syrian government daily, Al-Thawra, said. Another possibility given was that the virus was created to attack "the yellow race - especially in China and Vietnam" that are "rising powers" threatening "American hegemony over the world."

Introducing an Iraqi Patriot

March 29, 2006

Last week in the Washington Post, Fareed Zakaria of Newsweek authored a column titled, "Why Iraq Is Still Worth the Effort." He wrote, "Iraq has no Nelson Mandelas, but many of its leaders have shown remarkable patience, courage, and statesmanship...." One such leader is Sayyed Raouf Ayad Jamal al-Din. He is the Iraqiya list's representative for the predominantly Shiite province of Nasiriyah. Led by a former prime minister, Ayad Allawi, this political grouping is the major progressive movement in post-Saddam Iraq made up of both Sunni and Shiites that are secular, liberal, and supportive of women's rights.

Mr. al-Din frequently conveys support for American intervention in Iraq:

* "We must discipline ourselves before the Americans ... discipline us. We must maintain human rights, which we have neglected for 1,300 or 1,400 years to this day, until the arrival of the Americans. ... They came to teach you, the followers of Muhammad, how to respect human rights." - LBC TV, July 31, 2005

* "We are grateful to America for ridding us of school curricula which promote hatred and animosity." - Al-Arabiya TV, October 10, 2005

Mr. al-Din is a vocal critic of those he refers to as the "scum" destroying his country and is fearless in taking on Arabs who obstruct progress in Iraq:

* "The Arabs must stop meddling in Iraq's affairs, and stop inciting hatred, violence, and terrorism. They should not call these terrorist attacks 'resistance.' ... They are scum. ... We have not heard a single Arab jurisprudent condemn the terrorist attacks." - LBC TV, July 31, 2005

Speaking on Al-Arabiya TV on January 3, 2005, Mr. al-Din was also highly critical of Iranian government "interference" in Iraq accusing that country of trying to add "despair and frustration" among the lives of Iraqis. In response to any suggestions that the average Shiite Iraqi is pro-Iranian, he stated: "I tell the Iranians that if they hear that pictures of Ayatollah Khamenei are hung in Iraqi homes, it is a lie."

During a May 15th interview with Lebanon's LBC TV, when asked his opinion about how to deal with terrorists in Iraq, he referred to them as "the dirtiest type of riffraff among the Arabs," and explained: "In my opinion, it is not permissible to talk to these murderers. A decisive military action is the only remedy."

Mr. al-Din is also a fierce opponent of the Arab ideology of hatred which has bred Al-Qaeda:

* "Who turned Al-Zarqawi into a criminal murderer? ... He was just a little boy crawling into his mother's arms, but then, out of his love for the religion and the mosque. ... the imam got hold of him, and turned him into what he is." - Al-Fayhaa TV, November 30, 2005

Mr. al-Din is highly critical of Arab states for their unwillingness to support his country, stating they fear the spread of democracy:

* "The terrified and self-defeated Arab states, who fear the establishment of a democratic regime in Iraq, would prefer a stupid and reckless dictator like Saddam to a democratic regime in Iraq." - LBC TV, July 31, 2005

* "How come this country [Israel] has developed a democratic regime, although it too has been at war and is surrounded by enemies, whereas the Arabs have not developed democratic regimes, using the existence of Israel as a pretext?" - Al-Arabiya TV, January 3, 2005

At the age of 16, Mr. al-Din demonstrated against Saddam Hussein's suppression of Iraqi Shiites practicing their religion. After becoming a target of Saddam's secret service, he fled to Syria, to Iran for eight years of religious studies, and then became leader of the Shiite community in Dubai.

His vision for the new Iraq includes a country that allows freedom of worship of all religions. An April 17, 2003, Financial Times article reported on an early meeting of leading Iraqis in the post-Saddam era, "Sheikh Ayad Jamal Al-Din ... surprised many by demanding 'a system of government that separates belief from politics. Dictators may not speak in the name of religion,' he said quoting the Koran."

Similarly, he stated this summer:

* "I support a secular regime, which will fully separate religion and state. ... I believe that my freedom as a Shiite and as a religious person will never be complete unless I preserve the freedom of the Sunni, Christian, Jew, Sabai, or Yazidi." - LBC TV, July 31, 2005

Mr. al-Din will be visiting Washington D.C. next week. He is an important visionary who is on the forefront of building the new Iraq. He is precisely the kind of man Iraq now needs to lead it and can likewise serve as an inspiration to the entire Arab and Muslim world.

How Hamas Greeted 9/11

April 5, 2006

Hamas praised the killing of thousands of Americans in the attacks of September 11, 2001, while saying Muslims could not have been involved - a similar response to those of other groups in the Middle East. Now that Hamas is running the Palestinian Authority, it is worth taking a second look at the terrorist organization's official press outlets' coverage of the attacks.

* "Allah has answered our prayers," Dr. Atallah Abu Al-Subh wrote in an open letter titled "To America," which appeared September 13, 2001, in the Hamas mouthpiece Al-Risala. The Hamas publication also directed messages to high-ranking members of the Bush administration:

* "Every time Dick Cheney and his girlfriend Condoleezza Rice ... admonish us [and] gloat at our misfortune, they incite more [violence against us]! We stand in line and beg Allah to give you to drink from the cup of humiliation - and behold, heaven has answere. ... America, you planted in the hearts of all men ... the seedling of hatred of you! You never considered that the day would come when the saplings would grow and put out your eyes ... at the top of the World Trade Center ...Those saplings grew and spread and struck at the liver of the Pentagon. ... Don't you see that thorns have reached the eyes of your strong Secretary of Defense, Rumsfeld, who thought that he was too strong to be hurt by the thorns of vengeance ... Perhaps you realize, after Rumsfeld fled, that he is only an ordinary man," Al-Risala contributors wrote shortly after the attacks.

* "The Jews infiltrated the American Army, particularly the Air Force and they pressured pilots to take the planes, knowing that religion is not denoted in the identity cards of those joining the American military. Accordingly the airplanes were [controlled] by the Jews. ... Why is Bush ignoring these facts.... Don't act like you don't know who is behind the recent events..." a Hamas activist, Yussef Al-'Azam, wrote in a Hamas newspaper, Al-Sabil, on October 4, 2001.

* "Should I greet you [anthrax], or should I curse you ... I will begin by saying:

Oh Anthrax ... you have sown horror in the heart of the lady of arrogance ... you have entered the most fortified of places; [you have entered] the White House and they left it like horrified mice," a November 1, 2001, editorial in Al-Risala headlined, "To Anthrax," said. The letter also belittled the American military and concluded by blessing the attacks on America:

* "The Pentagon was a monster before you entered its corridors. ... And behold, it now transpires that is men are of paper and its commanders are of cardboard. ... This horror that you have sown ... in the heart of the bloodsucker [the America] makes me think as I do. ... Our hearts ... were filled with belief that Allah is capable of defeating America by means of the weakest of his earthly soldiers, after he used you to sow horror in their hearts. ... The Americans see you as an imminent attack that is about to shake the lady with the proboscises that suck the blood of ... peoples ... 'Allah bless you,'" "To Anthrax" continued.
* "How much hatred has [America] stockpiled. ... Were the eradication of its Marines in Lebanon, the destruction of its military headquarters at Khobar in Saudi Arabia, the destruction of the USS Cole in Yemen, the bombing of its embassies in Zambia [sic] and in Kenya, and the attacks on its soldiers in the Gulf ... not sufficient? The U.S. should have learned the lessons of history ... Today, the U.S. has treated the bitter harvest it sowed in the hearts of millions. Today, we ask: After this, will the U.S. [continue to] think that it is above the law?" the editor of Al-Risala, Dr. Ghazi Hamad, said in justification of the attacks days after September 11, 2001.

Some of the most vulgar responses to the attacks of September 11, 2001, were the celebrations in the Palestinian streets, including shouts of joy by women. Hamas' reaction to the attacks was equally as disturbing yet not really shocking.

Passover and the Blood Libel

April 12, 2006

The embrace by prominent figures from the Middle East - including members of the Saudi royal family and a former Syrian defense minister - of blood libel over the past few years has been well-documented.The most influential government-controlled newspapers in Egypt, Syria, Saudi Arabia, Iran, and elsewhere similarly have promoted the claim that Jews incorporate blood from Muslim and Christian children in Passover matzo.

The blood libel tale made headlines again this year after MEMRI reported that a new Jordanian TV channel, Al-Mammou, began airing the series "Al-Shatat" during Ramadan. "Al-Shatat is a Syrian-produced, Hezbollah-backed, Al-Manar TV series that was first shown in 2003 and includes scenes of Jewish rabbis kidnapping Muslim children to use their blood for "tasty" Passover matzo. Since 2003, the program has been rebroadcast throughout the Muslim world, in countries such as Lebanon and on Iranian TV channels 1 and 2 (in 2004) and Sahar TV (in 2005).

After severe criticism, the Jordanian Embassy in Washington, D.C., sent a press release to notify MEMRI that the series had been pulled. This unprecedented move by an Arab country should be applauded.

Unfortunately, the spread of the blood libel theory from Jordan was not a one-time event. The secretary-general of the Islamic Action Front Party in Jordan, Sheikh Hamza Mansour, appeared on Saudi Al-Majd TV on March 22, 2004, and said, "As the Koran says, You shall find the bitterest of people in enmity against the believers to be the Jews ... This is their sick psychological nature. This is the nature that makes matzo out of innocent children's blood."

A Jordanian-produced series titled, "Stories From the Koran," which aired on Saudi Iqra TV in February 2005 included a scene about how Jews "altered" the Torah and discussed the Jewish "trait" of hating. A Jewish character, Huyay, recited the following dialogue: "We are the slayers of prophets, and we live off their blood! ... Once they are dead, we finish off their followers." Jewish characters in the series also played with voodoo dolls of Islam's Prophet Muhammad.

Other notable blood libels from the Arab and Iranian press include:

* "The Talmud, the second holiest book for the Jews, determines that the 'matzos' of Atonement Day [sic.] must be kneaded 'with blood' from a non-Jew. The preference is for the blood of youths after raping them!!!" Dr. Mahmoud Al-Said Al-Kurdi wrote in Al-Akbar of Egypt on March 25, 2001.

* "Unfortunately, the West has forgotten two horrendous incidents, carried out by the Jews in 19th-century Europe ... In 1883, about 150 French children were murdered in a horrible way in the suburbs of Paris, before the Jewish Passover holiday. Later research showed that the Jews had killed them and taken their blood ... A similar incident took place in London, when many English children were killed by Jewish rabbis." - From a December 20, 2005, discussion on Iran's Jaam-e-Jam 2 TV about denial of crematoriums at Auschwitz, in which Iranian political analyst Dr. Hasan Hanizadeh also discussed the subject of Passover.

* "The bestial drive to knead Passover matzos with the blood of non-Jews is [confirmed] in the records of the Palestinian police, where there are many recorded cases of the bodies of Arab children who had disappeared being found torn to pieces without a single drop of blood. The most reasonable explanation is that the blood was taken to be kneaded into the dough of extremist Jews to be used in matzos to be devoured during Passover," Adel Hamooda of Al-Ahram in Egypt wrote October 28, 2000.

* "I would like to clarify that the Jews' spilling human blood to prepare pastry for their holidays is a well-established fact ... How is it done? For this holiday, the victim must be a ... Christian or Muslim. His blood is taken and fired into granules. The cleric blends these granules into the pastry dough ... For the Passover slaughtering ... the blood of Christian and Muslim children under the age of 10 must be used, and the cleric can mix the blood [into the dough]..." Dr. Umayma Ahmad Al-Jalahma of King Fahd University said in Saudi Arabia's Al-Riyadh in March 2002.

* "The Jews slaughter non-Jews, draining their blood, and using it for Talmudic religious rituals," Hussam Wahba, a columnist for the religious Egyptian magazine Aqidati - published by a foundation linked to President Mubarak's National Democratic Party - wrote in a August 2004 article.

Sadly, these are just a few of the countless examples of blood libels emanating from the Middle East press. To learn more, see MEMRI's Antisemitism Documentation Project (http://memri.org/antisemitism.html).

Time for All Muslims To Condemn Iran's Holocaust Denial

April 19, 2006

> *"We must make sure that their deaths have posthumous meaning. We must make sure that from now until the end of days all humankind stares this evil in the face ... and only then can we be sure it will never arise again"*
>
> — PRESIDENT REAGAN, DURING REMARKS GIVEN AT THE SITE OF THE U.S.HOLOCAUST MEMORIAL MUSEUM, OCTOBER 5, 1998.

Since President Ahmadinejad came to power in Iran, he has become known for three things: pursuing Iran's nuclear program; promising to destroy Israel; and denying the Holocaust.

To the shock of the civilized world, Mr. Ahmadinejad first made headlines denying the Holocaust on December 14 — and proceeded to repeat the sentiment at later dates. "They have invented a myth that Jews were massacred," he said.

An Iranian affairs specialist, Ali Nourizadeh, appeared on Abu Dhabi TV on December 24 to explain what motivated Mr. Ahmadinejad's speeches. Calling his words "no slip of the tongue," Mr. Nourizadeh said the Iranian leader is working to coax America and Israel to attack Iran "because Ahmadinejad and his gang need a new 'Karbala' in Bushehr."

Mr. Nourizadeh warned that Mr. Ahmadinejad and his cohorts want to create "a new legend" to incite Iranian zealots who "believe that their war [with the West] is coming, and that the Hidden Imam [Messiah] will appear after this war.This is the backward belief that Ahmadinejad defends."

Since he was elected last summer, there has been an unprecedented wave of Holocaust denials coming out of Iran, including those in the form of government-sponsored symposia and cartoon contests.

Recently, Western Holocaust deniers also have been quoted extensively and celebrated in the Iranian press, including David Irving; FredrickToben of the Adelaide Institute in Australia; professor Robert Faurisson of France; the director of Canada's Association for Free Expression, Paul Fromm, and Roger Garaudy, who, on December 13, said on Iran's TV Channel 1: "No Jews or other prisoners were killed by poisonous gas — not in Dachau, Bergen, or Buchenwald."

In recent months, American Holocaust deniers have also been covered by the Iranian press, including a California-based Saudi professor, Dr. Abdullah Muhammad Sindi; a Northwestern University professor, Arthur R. Butz, and a former Associated Press writer, Michael Hoffman II. The following are examples of homegrown Holocaust denial from the Iranian press:

- "If the issue of the Holocaust was true, Europeans would not be afraid of investigating about it." The Majilis speaker, Gholam-Ali Haddad Adel, IRNA, April 16.
- "Following World War II, the Jews ...tried to present themselves as oppressed ... They invented a false claim.They said that Hitler burned 6 million Jews in crematoria. They tried very hard to convince the world that this issue, this lie, was true, and they succeeded." Ayatollah Ali Meshkini, Iranian Channel 2, December 17.
- "If 6 million people had died, wouldn't there be a record somewhere of at least a hundred names?" Dr. Majid Goudarzi, Iranian TV Channel 2, January 5.
- "The killing of 6 million Jews in the crematoria is a myth, not a fact ... For hygienic reasons, they used to burn the bodies of those who died of ... contagious disease ... this means the crematoria were used for hygienic, not political purposes ... the gas chambers were used for disinfecting the clothes." Dr. Hosein Rouyvaran, an Iranian political analyst, IRINN TV, December 27.
- "Historians and eyewitnesses ... who saw with their eyes what happened 70–80 years ago ... deny the existence of crematoria at a place called Auschwitz . . . a case of burning people has been registered in history ... the people who were burnt were in fact Christians ... They were burnt by ... Jews." Dr. Ali-Reza Akbari, Jaam-e-Jam 2 TV, December 20.
- "The rooms where gas was supposedly injected from above had no holes for gas injection ... No holes, no Holocaust." Amir Mohebbian, an Iranian political analyst, IRINN TV, January 3.

A spokesman for Germany's oldest Muslim organization, the Islamic Archiv-Deutschland Central Institute, responded to Mr. Ahmadinejad's Holocaust denial in February, calling it a "disgrace to all Muslims." The spokesman went on to invite the Iranian leader to tour the Auschwitz-Birkenau concentration camp with his organization, and challenged Mr. Ahmadinejad, "In this place of horror he can again deny the Holocaust, if he has the courage."

Sadly, the statement against Mr. Ahmadinejad and his Holocaust denial is an exception. As Tuesday marks Holocaust Remembrance Day, one can only hope other voices from the Muslim world will follow.

Freeing a Parliamentarian

April 26, 2006

A former Egyptian parliamentarian and leader of Al-Ghad Party (the Party of Tomorrow), Dr. Ayman Nour, became an international cause celebre following his arrest last year by the Egyptian government for forging signatures on the petition to secure the formation of his new party - an allegation he has denied vehemently. He has been locked in a prison cell for the last 142 days or 3,408 hours.

According to freeaymannour.org, Al-Ghad stands for liberal and democratic values, with a focus on human rights. Other pillars of the party are encouraging constitutional reform, limiting presidential powers, and creating open, multi-candidate presidential elections in Egypt.

Freeaymannour.org offers up-to-date articles from the international press regarding Dr. Nour's case, as well as information from Western legislators who have spoken out on his behalf. For example, the site highlights U.S. House of Representatives Resolution 82, from March 3, 2005, which "express[ed] the grave concern of Congress" regarding Dr. Nour's arrest. A December 24 statement from the White House, which called Dr. Nour's five year prison sentence "deeply troubling," is also posted at freeaymannour.org.

The following are some examples of the dozens of personal letters and emails in support of Dr. Nour, from influential Europeans who also have been outraged by his imprisonment and are active in working to secure his release:

* A text adopted by the European Parliament, "Urges the Egyptian authorities not to undermine the prospects recently opened ... to multi-candidate presidential elections, and to press ahead with democratic reforms; expresses its deep concern ... about the conviction of Ayman Nour ... regards this as a seriously retrograde step and calls on the Egyptian authorities to make every effort to ensure that this case is correctly dealt with." From a letter sent to freeaymannour.org by Czech politician Jana Hybaskova, member of the European Parliament, on March 8.

* [We the European Parliament] welcome and support the worldwide calls to release Ayman Nour and strongly urge the Egyptian authorities to ensure that

Ayman Nour is well treated and not sub ject to torture or other ill treatment and to further ensure that he is given prompt, regular and unrestricted access to his lawyers, doctors (as he's diabetic) and family." A January 18 e-mail from an Italian politician and member of the European Parliament, Emma Bonino.

Most importantly, freeaymannour.org provides visitors an opportunity to sign a petition addressed to President Mubarak of Egypt urging Dr. Nour's release. It calls the prison sentence "an attack on the nascent initiative to a free political life in Egypt to move towards democracy and pluralism."

The petition asks the international community to support Dr. Nour and the struggle for a free and democratic Egypt. It states that all individuals and associations who care for the future of Egypt and of the Arab and Muslim world as a whole to add their signature.

To date there have been more than 600 signers including Dr. Nour's wife Gameela Ismail; Arab reformers, such as Saad Eddin Ibrahim; 27 members of the European Parliament from Finland, Britain, Italy, Sweden, the Czech Republic, Poland, France, and Germany; members of academia; journalists; activists; researchers; students; housewives; lawyers, and concerned citizens of the world.

The signers are from countries such as Egypt, Morocco, Iraq, Jordan, Tunisia, the Palestinian Territories, Algeria, Syria, Lebanon, Saudi Arabia, Kuwait, U.A.E., Libya, Spain, Switzerland, Thailand, America, Canada, Brazil, the Netherlands, Austria, Norway, and the Ivory Coast.

The Web site also requests visitors to "officially" ask Egyptian authorities to visit Dr. Nour in prison to help ensure his safety and to direct such requests to: 1. Habib el Adly, Egyptian minister of Interior, Lazoghly, Cairo, Egypt, Fax no. 202-795-7286. 2. General Mahmoud Wagdy, Head of the Egyptian prisons, Kolaly, Cairo, Egypt.

For more information or to sign the petition visit www.FreeAymannour.org or e-mail webmaster@freeaymannour.org.

For more information on Arab and Iranian Reformists visit http://memri.org/reform.html.

Bin Laden's Challengers

May 3, 2006

"Al-Qaeda is not the organization it was four years ago," the State Department said last week in its annual report on Global Terrorism. Commentators in the Arab press, too, in the days following the April 23 release of his latest tape, have observed a wounded Osama bin Laden.

"He sounds deeply disappointed that his messages have fallen on deaf ears. The caravan of jihad has left him behind," a Middle East analyst and Sarah Lawrence College professor, Fawaz Gerges, told Al-Jazeera, describing the Al-Qaeda leader as "desperate."

"Bin Laden may continue to remind the world of his continued existence thanks to Al-Jazeera. But his time has come and gone," Amir Taheri similarly wrote in the April 28 edition of a London-based Arabic daily, Al-Sharq Al-Awsat.

Al-Sharq Al-Awsat's editor in chief, Tariq Alhomayed, said in an analysis of the bin Laden missive, "The most dangerous part of his speech was clear as he referred to combating Arab liberal intellects who he describes as 'the mockers of Islam.'" Mr. Alhomayed noted that both the Western and Middle Eastern press avoided discussing this part of the speech, and explained that by Mr. bin Laden's definition, a "liberal" is anyone who disagrees with his ideology.

According to Mr. bin Laden the world is divided into two camps: Those who follow the ideology of Al-Qaeda and everyone else. Now members of the latter, including Arab intellectuals, are lining up to denounce the terrorist organization's leader and challenge his ideology.

"Why has there been no fatwa against bin Laden?" a Jordanian intellectual who lives in America, Dr. Shaker Al-Nabulsi, asked in an August 7 article in a Kuwaiti daily, Al-Siyasa.

"Despite the fact bin Laden murdered thousands of innocents in the name of our religion and despite the damage he has caused to Muslims everywhere ... not a single fatwa has been issued calling for the killing of bin Laden," a former Kuwaiti minister of information, Dr. Sa'd Bin Tefla, wrote in Al-Sharq Al-Awsat on August 30, 2004.

A moderate Islamic scholar, Sheikh Abd Al-Muhsin Al-'Abikan, who serves as an adviser to the Saudi Justice Ministry and a member of the Saudi Shura Council, answered the call May 15. He expressed a willingness to confront Mr. bin Laden and his supporters publicly.

"Those who believe in the ideology of Al-Qaeda have little knowledge of Islam. So I have challenged them to a confrontation ... Yes, I will publicly confront bin Laden ... I am willing to confront anyone who defends the ideas of Al-Qaeda, at any time," Al-Sharq Al-Awsat quoted Sheikh Al-'Abikan as saying.

Following the summer bombings at Sharm Al-Sheikh, an Egyptian scholar and columnist, Mamoun Fandy, also challenged Al-Qaeda and called upon Muslims to issue religious edicts against terrorism: "Just as bin Laden and his group describe moderate Muslims as followers of the West and as unbelievers, it is time for the Muslim leaders to proclaim bin Laden himself to be an unbeliever."

Even before the attacks of September 11, 2001, a former dean of the faculty of Islamic law at the University of Qatar, Dr. Abd Al-Hamid Al-Ansari, said the Muslim world needed to launch an ideological, cultural, educational, informational, and religious campaign to eradicate terrorism emanating from the Arab world. The text of one article Dr. Al-Ansari wrote in April 2004 read, "the Muslim public has been deceived" by Mr. bin Laden and his jihad.

In an interview with Bahrain TV on March 2, 2005, Dr. Al-Ansari proposed a solution for dealing with Mr. bin Laden, his supporters, and their ideology of hate: "After 50 years of enlightenment, education, and modernization, along comes someone like bin Laden ... We need to act as Germany upon its defeat. The German elite reexamined its culture and found that the entire German culture is full of poison. Its sources are full of poison."

As the war on terror continues, the West must support and invest in true Arab reformists, weeding out self-proclaimed ones - often found quoted in leading Western press outlets.

Tariq Al-Homayed described such impostors in a November article in Al-Sharq Al-Awsat: "We suffer today from [double] talk, from those who to CNN will brand Osama bin Laden a 'terrorist,' yet will appear on Al-Jazeera describing him as 'Sheikh Osama.'"

It could be interpreted by reading Mr. bin Laden's most recent speech that the ongoing impact of Arab reformists has now become one of his major preoccupations.

Bin Laden's Challengers, Round II

May 10, 2006

In Osama bin Laden's April 23 audiocassette, he attacked Arab "freethinkers," naming several and calling for their deaths. As a precedent, the Al-Qaeda leader referenced Islam's Prophet Muhammad, who killed Ka'b ibn Al-Ashraf for writing defamatory poetry. Mr. bin Laden then instructed his followers: "It is intolerable and outrageous that the heretics are among us, scorning our religion and our Prophet ... Do not consult anyone about the killing of these heretics. Be secretive in carrying out that which is required of you."

One of the "heretics" Mr. bin Laden spoke of on his most recent tape is a Kuwaiti intellectual and University of Kuwait political science professor, Ahmad Al-Baghdadi. In the years leading up to the attacks of September 11, 2001, Mr. Al-Baghdadi was an outspoken critic of Islamism and, in a March 16 article, referred to Mr. bin Laden as a vile terrorist and murderer.

Following September 11, Mr. Al-Baghdadi penned an article about the attacks that was printed in many Arabic newspapers, including Akhbar Al-Youm of Egypt.

"The Palestinian Arabs were the first to invent airplane hijacking and the scaring of passengers ... Arab Muslims have no rivals in this; they are the masters of terrorism towards their citizens, and sometimes their terrorism also reaches the innocent people of the world, with the support of some of the clerics. Today, the Arabs and the Muslims are paying the price of their terrorism ...They are persecuted and humiliated across the civilized world. They are rejected in both the West and the East. In restaurants, in airplanes, in buses - everywhere they are spit upon. One cannot complain to the West for what it is doing to them, because [in] the Arab and Muslim world, everyone ... [is] lying about terrorism ... The Islamic world ... [is] the only [place] in which intellectuals - whose only crime was to write - rot in prison. The Arabs and the Muslims claim that their religion is a religion of tolerance, but they show no tolerance for those who oppose their opinions," Mr. al-Baghdadi wrote.

The Kuwaiti government on multiple occasions has punished Mr. Al-Baghdadi for expressing his opinions. He was first jailed in 1996, and in 1999 was given a one-month "suspended" sentence for "insulting Islam" in an article in a University of Kuwait student magazine. More recently, he was given a one year "suspended" sentence following the publication of an article in June, 2004, that expressed critical views of religious education in Kuwait. Specifically, he said schools teach children "how to disrespect women and non-Muslims."

The political attacks on Mr. Al-Baghdadi culminated in March, 2005. After a Kuwaiti court condemned him to another one-year prison sentence, he published a public request for political asylum in the West in which he observed, "a morbid atmosphere full of bacteria and viruses of hatred and tyranny."

The content of a series of articles by Mr. Al-Baghdadi that were published November 14-17, 2004, in the Kuwaiti daily Al-Siyassa, help to explain why he was singled out by Mr. bin Laden. The professor criticized "Islamic religious thought" for being misogynistic and intolerant, and said it presents obstacles to freedom of thought and speech and human health in matters of treatment and medicine. The professor went on to say such thinking supports political tyranny by oppressing democracy. "Proof of this are the supplications and appeals [to Allah] that we hear in the mosques to destroy all non-Muslims," he wrote.

In the series he praised America and Europe for "saving many peoples from oppression, while sacrificing [their own] human life and property" and criticized Muslim leaders for failing to aid non-Muslims. When contrasting Muslims living in the West to non-Muslims living in the Muslim world, Mr. Al-Baghdadi said, "There is no Islamic country in which a Christian or a Jew could reveal a cross or a skullcap and get away with it peacefully ... without people harming them." He also praised the West for supporting freedom of expression, which, he said, is lacking in the Muslim world.

He went on to note that the government must approve the construction of churches in Muslim countries, and that no Western countries prohibit the building of mosques or prevents Muslims from praying in public. "There is no church in the secular Christian world in which a priest stands and curses anyone who disagrees with his religion or prays for trouble and disaster to befall them, as do the preachers in our Friday sermons," Mr. Al-Baghdadi said.

Mr. bin Laden's statement of April 23 also called for the killing Turki Al-Hamad. Next week's column will focus on this Saudi reformist.

Bin Laden's Challengers, Round III

May 17, 2006

In Osama bin Laden's April 23 tape, he called for the killing of prominent Arab reformers, including a Saudi author, Turki Al-Hamad. This was not the first time threats were made against Mr. Al-Hamad. Saudi religious figures have issued four fatwas calling for his death. He told the British Broadcasting Corporation's "Hardtalk" in 2004 that his family members, too, have received death threats by phone and e-mail. Official Al-Qaeda communiques have also threatened "the apostate."

Mr. Al-Hamad is best known for a trilogy of coming-of-age novels considered dismissive of prominent Islamic works. First came "Adama," named after a neighborhood in Damman, Saudi Arabia. Following its 1998 release, the book quickly sold 20,000 copies before being condemned as "heretical" and banned in Saudi Arabia, Kuwait, and Bahrain. The London publishing house Saqi Books, which translates Mr. Al-Hamad's writings, hailed it as "the most explosive novel to emerge in the Middle East in years."

On the dust jacket of his second novel, "Shumaisi," named for a quarter of Riyadh, Mr. Al-Hamad wrote: "Where I live there are three taboos: religion, politics, and sex. It is forbidden to speak about these. I wrote this ... to get things moving." His third novel, "Karadib," named after a prison in Jeddah, also focused on social taboos of conservative Middle Easterners.

In the summer of 2005, Mr. Al-Hamad released a novel loosely based on the attacks of September 11, 2001, and four of the hijackers titled, "The Winds of Paradise." The New York Times quoted him as saying, "I wrote the latest book just to say that the problem is not from outside [Saudi Arabia], the problem is from ourselves - if we don't change ourselves, nothing will change."

In a 2003 supplement from the Saudi English daily Arab News about September 11, 2001, Mr. Al-Hamad wrote that the attacks have "not incited any crucial questions among Arabs." He called the response "mostly a mixture of justifications and excuses," which, he added, "try as usual for the Arabs to blame the others and to exonerate themselves."

Mr. Al-Hamad is an outspoken critic of Saudi government curricula, universities, and summer camps, and he has connected the September 11 attacks to the Saudi educational system. On August 25, 2004, Mr. Al-Hamad appeared on Al-Arabiya TV, saying, "We must be wary even of official [Saudi] institutions, such as summer camps, [and] Koran memorization schools."

Of the September 11 suicide bombers, he said, "When you study their biographies ... they are graduates of official institutions, universities, official centers, and so on ... I say, pay attention to what's happening inside these summer camps, inside these universities, and colleges ... A politicized religious ideology is being indoctrinated into the students."

In 1995 at the age of 45, Mr. Al-Hamad retired from King Saud University. In interviews, he said officials were censoring what he could teach, and that he could better influence young Saudis by writing novels. Mr. Al-Hamad told the New York Times in 2005: "The first thing [to bring democracy] is that you have to use the educational system to spread different values, human values." He also decried the fact that students quote Islamist religious figures such as Sheikh Ibn Taymiyyah to explain scientific theories.

In an interview with Germany's Qantara.de, Dialogue With the Islamic World, in August 2004, Mr. Al-Hamad again discussed the Saudi curriculum: "Certain values must be spread by our education system, for example tolerance, getting along with others as human beings, as other races, or as 'nonbelievers.' There are values that must be planted inside the minds of young children if we are to have a new society. If we continue teaching our children hatred and intolerance, then these ills will grow and a climate for terrorism will be created," he said.

Turki Al-Hamad, Ahmad Al-Baghdadi, and other Arab reformists mentioned in this series are not alone in the fight against Al-Qaeda and the ideology of hate. Others whom Osama bin Laden and his followers view as "apostates" are emerging as a rival force.

As an Arab reformist who now resides in America, Dr. Shaker Al-Nabulsi, explained on May 1 to the Arabic Web site Mid East Transparent, "The forces of oppression and darkness draw back every day while the forces of light and enlightenment advance every day. However, they advance slowly ... since the forces of darkness perform daily barbaric acts of terror and publicized these acts on the various media channels, and consequently they spread more quickly than the forces of light that have no weapons except words, and no sword but the sword of dialogue."

The Meaning of 'Islamofascism'

May 24, 2006

Since the war on terrorism began many new Arabic and Islamic terms including "Fatwa" and "Jihad" have been introduced to Western vernacular. To the displeasure of some Islamists other terms such as "homicide bomber" and "Islamofascism" have also been introduced.

An article titled "How the Right Played the Fascism Card Against Islam" on the Web site of the Council on American-Islamic Relations is highly critical of using the term "Islamofascism":

"Fascism is coming back into fashion, at least in the propaganda wars. For the right, it comes in the shape of a new word: 'Islamofascism'. That conflates (sic) all the elements into one image: suicide bombs, kidnappings, and the Qur'an; the fighting in Iraq and Afghanistan; Iranian clerics and Hitler ... Usage has gathered momentum among commentators and academics who seek a verbal missile to debilitate those who disagree with them. They have adopted it as a sort of Judeo-Christian war cry."

Unlike CAIR, many Arab liberal and reformist writers, have supported such an analogy.A Saudi columnist, Muhammad bin 'Abd Al-Latif Aal Al-Sheikh, published a series of articles in the Saudi daily Al-Jazirah in July, attacking the ideology of the Al-Salafiyya movement (long associated with Saudi Wahhabi Islam). He said that the ideology of this movement was similar to, or even worse than, the Nazi ideology, and that it should be dealt accordingly.

Mr. Al-Sheikh cited the conference on "de-Nazification" held in Potsdam, Germany, shortly following Nazi Germany's surrender in 1945 to serve as precedence today. He explained it was credited with uprooting the culture of Nazism from Europe and that the conference made Nazism into something similar to a crime, not just in judicial and political terms, but also in terms of culture, ideology, and the media.

Just as the world uprooted Nazism at the Potsdam conference, Al-Sheikh explained that the West should follow suit with Islamism after the attacks of September 11, 2001: "I still believe that one of the primary missions of the international community today is to repeat its experience with Nazism and to deal with this dangerous barbarian culture [of Islamofascism] exactly as it dealt with the Nazi culture. If this does not happen, the near future is liable to bring many [events], the consequences of which will be far more severe for all of humanity than [the consequences] of World War II."

A columnist for the London Arabic daily Al-Hayat, Zuheir Abdullah, blamed what he termed "Arab fascism" and "Islamism" for leading to the current backwardness of the Middle East. In an August 2003 article he wrote, "since 1948, primitive Arab fascism," sometimes "allied with fundamentalist Islam," has produced only "empty slogans." Mr. Abdullah continued, "many simple-minded people and ignorant persons were unfortunately brainwashed and turned into the fuel of this extremism." He concluded by stating the Arab world's embrace of fascism and Islamism has led it to adding "almost nothing" to modern civilization.

In a liberal Arabic Web site, elaph.com, on December 13, an Iraqi reformist, Dr. Abd Al-Khaleq Hussein, explained that the Iranian and Syrian regimes are fascist in the true sense of the word, and that no stability, democracy, or economic prosperity is possible in the region as long as they exist.

Dr. Hussein said, "Some believe that describing the Iranian Islamic regime and the Baath Party's Syrian regime as fascist is offensive ... I call a spade a spade, in accordance with the scientific definition of the terms 'Fascism' and 'Nazism.' Italian Fascism, like German Nazism, was a racist movement that discriminated among human beings on the basis of their racial or national affiliation ... The common denominator defining an ideology or policy as Fascist or Nazi is discrimination among human beings ... Since both the Iranian and Syrian regimes discriminate among the various groups within their peoples on the basis of religion and ethnicity, they are no different from any other fascist or Nazi regime."

Dr. Hussein is hopeful that the West will treat "Islamofascism" in the Middle East as Nazism in Europe was treated following World War 2 and that it will lead to similar results: "We are optimistic about the future ... The first domino has already fallen, when the fascist Baath Party fell in Iraq.The rest of the dominoes will follow. Just as Nazism and Fascism fell in Europe, and the rest of the totalitarian, tyrannical regimes followed, it is inevitable that history will grind down the remaining fascist regimes [in the Middle East]."

The Darfur Conspiracy Revisited

May 31, 2006

In "The Darfur Conspiracy," this column on October 19 addressed conspiracies from the Arab and Iranian press about the genocide in Sudan. Such press outlets claimed that the violence resulted from American attempts to control Sudanese oil, plans to create a Christian state in Sudan, and Jews trying to annex the African country to Israel.

"Look for the Zionists behind every disaster. We have found their fingers in Darfur," one of the most influential Islamists, Sheikh Yosef Al-Qaradawi, said in an interview with a Qatar daily, Al-Sharq, on September 10, 2004. Similarly, before the last American presidential election, Sudan's ambassador in Tehran told Iran Daily, "The Zionists and rightist Christians are fanning rumors against Sudan."

The last six months have seen countless efforts by the West to stop the genocide in Darfur, including activity at the United Nations, rallies in Western capitals, and numerous newspaper and television programs devoted to the killings. April 30's Save Darfur rally on the National Mall boasted politicians, celebrities, religious leaders, human rights activists, and Darfur refugees, united in their call for action.

Meanwhile, many in the Arab and Muslim world, including Al-Qaeda, Iran's state-controlled press, and mainstream Arab and Muslim American leaders, continued to embrace conspiracy theories surrounding Darfur.

In a May 8 column titled "Why I Spoke at the Darfur Rally," the president of the Arab American Institute, James Zogby, said his decision to participate was "complicated" due to "the make up of the U.S.-based movement" that sponsored the event.

"It is a fact that a number of Evangelical Christian organizations who had been engaged in controversial missionary/conversion efforts in Darfur were involved, as were some Jewish groups who had a history of using Sudan as an issue to drive a wedge between Arabs and Africans," Mr. Zogby said.

In a May 16 article titled "Condemning Zionism Is Crucial to World Peace, and Essential to Peace in the Muslim and Arab World," a board member of the Council on American Islamic Relations, Anisa Abdel Fattah, wrote about "possible attempts [by Zionists] to expand into Sudan through Darfur."

A New Jersey-based Arabic Web site, Free Arab Voice, www.freearabvoice.org, published by and for Arab-Americans, recently posted an article by Abu Iskandar as-Sudani on its homepage. The article originally appeared in the May 2005 edition of a Palestinian Arab magazine, Al-Hadaf, and was devoted to Darfur. The author wrote that the claims of genocide were invented by the Western press, and discussed "the unfolding Zio-American plan to break up the Arab states along the lines of the strategy for a Greater Middle East." He concluded by saying this strategy "would give the American Administration and Israel control of [Darfur's] agricultural, petroleum and mineral resources of the country and facilitate their plunder of its uranium deposits." Free Arab Voice links to terrorist Web sites such as those of Hezbollah and Hamas.

Osama bin Laden's April 23 audiotape also claimed there was an American-led, Zionist-crusader conspiracy to separate southern Sudan and called on his mujahedeen in Sudan to be prepared for a long war against the crusaders.

A May 24 Islamic Republic News Agency article titled "Zionist Regime Uses 'Darfur Crisis' as Distraction" reported: "The powerful American Jewish lobby which tightly controls the American Congress" was behind "a high-profile propaganda campaign" about Darfur.

The communications director of the Muslim Canadian Congress, Tarek Fatah, wrote an article on May 10 titled "Why Are We Muslims So Silent on Darfur?" that appeared on the Web site of ArabAmericanNews.com and could be considered a response to the likes of Mr. Zogby and Ms. Fattah. In addition to discussing a Canadian rally to stop the genocide in Darfur, he was highly critical of Muslims for saying "the Jews have somehow stolen the issue of Darfur's genocide by actively campaigning against it."

"It is nonsense to suggest that the death, destruction, and all the suffering of the Darfurian people is imaginary or that Zionists are using it as propaganda ... To suggest that this is some sort of a U.S.-Israel conspiracy is ludicrous and insane," Mr. Fatah quoted El-Fadl El-Sharief, the Muslim Sudanese who organized the rally in Canada, as saying.

For information on how to help, go to www.savedarfur.org.

An Iranian Sermon Primer

June 7, 2006

Sermons airing on Iranian TV channels starring the supreme leader, Ayatollah Ali Khamenei; the secretary of the Guardian Council, Ayatollah Ahmad Jannati; the head of the Iranian Assembly of Experts, Ayatollah Ali Meshkini; the head of the Iranian Expediency Council, Ali Akbar Hashemi Rafsanjani, and a handful of other ayatollahs appear on any given Friday night in Tehran and on Saturday in Qom.

It should be noted that the words of the Iranian ayatollahs who run the country carry significant weight.

Responding to the sermons, frenzied crowds of Iranians chant, "Death to America," "Death to England," and "Death to Israel," with the occasional "Death to Saddam." Other popular chants include "America will be annihilated," "Islam will be victorious," "Woe to the enemy if Khamenei commands me to wage jihad," and countless other curses against the West.

The sermons focus on threats against the "Great Satan," America, and the "Little Satan," Israel, and their coming collapse. Other themes are attacks on President Bush, praise for jihad and martyrdom, Iraq, and support for Iran's nuclear program.

Below are excerpts from some of the more notable sermons over the past two years. They will be shown at a MEMRI lecture sponsored by members of the Senate next week on Capitol Hill.

AYATOLLAH JANNATI

"Western culture and civilization - and especially criminal America - are heading toward a serious collapse ..." - Channel 1, February 17, 2006

"The other possibility is that the English government itself caused this situation [the explosions in London]. This possibility was also raised regarding September 11 - that the Americans themselves caused it ..." - Channel 1, July 15, 2005

"These actions by our people will lead America to death ... and to eternal hell ..." - Channel 1, June 10, 2005

"Every Muslim and every honorable man who is not a Muslim must stand against the Americans, English, and Israelis, and endanger their interests wherever they may be ..." - Channel 1, June 4, 2004

AYATOLLAH MESHKINI

"The government of Libya ... surrendered to America ... the equipment of what it wanted to build [i.e. nuclear facilities] ... Gadhafi did not read the Koranic verse that says that they are the leaders of heresy ... we will never reconcile [with America] ..." - Channel 2, February 5, 2005

"My Lord, end the lives of three people shortly: Bush, Sharon, and Blair ... And the lives of their followers ..." - Channel 2, December 4, 2004

AYATOLLAH KASHANI

"If you [Americans] behave with disrespect ... [the Iranian people] will punch you in the mouth so hard that all your devouring teeth will fall off ..." - Channel 1, January 28, 2005

"I say to you, the American people ... your lives are lost ... America will collapse." - Channel 1, July 30, 2004

AYATOLLAH KHAMENEI

"The Americans should know: If they take any aggressive action against ... Iran, Iran will harm their interests wherever possible in the world ..." - Channel 1, April 26, 2006

"The freedom of speech about which the [Westerners] talk so much does not permit anyone to doubt the myth about the massacre of the Jews, known as the Holocaust ..." - Channel 1, February 7, 2006

"The American politicians, who are liars and hypocrites, conceal their satanic goals by accusing Iran of trying to make nuclear warheads ... When this nation's honor and this homeland's interests are at stake, we will sacrifice ourselves ..." - Channel 1, March 21, 2005

"Our people say 'Death to America,' and this is like saying, 'I seek God's refuge from the accursed Satan,' which is recited before any chapter of the Koran, even before 'In the name of Allah the compassionate, the merciful ..." - Channel 1, March 14, 2005

HASHEMI RAFSANJANI

"The biggest problem of the Islamic world today is global Zionism and the Great Satan, America. They are our enemies ... they have no religion and they do not accept God. ... Some are called Christians and some Jews ... every passing day the people's hatred festers, especially among the youth

> in Islamic countries. This hatred of the Great Satan and global Zionism constantly intensifies ..." - Kerman TV, April 17, 2005

> "We ... advise the distinguished people of Europe not to befoul themselves by tolerating the American pressure because eventually, they will fall into this trap and collapse like it." - Channel 2, June 19, 2004

MEMRI is the only organization in the world that tracks, documents, and provides excerpts of translated segments of Iranian sermons. Hundreds of hours of them are archived and can be viewed by visiting www.memritv.org.

Must-See Iran TV II

June 14, 2006

Senator Santorum, a Republican of Pennsylvania, and Senator Bayh, a Democrat of Indiana, will host a MEMRI-sponsored event tomorrow at the U.S. Capitol about Iranian government-controlled television.

Building on last summer's "Must-See Iran TV," MEMRI again has put together a critical set of translated clips from Iranian television programming that emphasize themes associated with Iran's nuclear aspirations, its acquisition of offensive weapons, anti-Americanism, and statements by leading Iranian government officials, including President Ahmadinejad.

The event will feature clips from the Iranian TV channels IRINN, Jaam-E-Jam 1-2-3-4, Al-Alam, Sahar, IRIB, and Iran 1-2-3-4, among others.

In addition, the briefing will highlight how the Iranian government has used its TV channels as a platform to disclose the latest developments of its military, therefore providing the West with critical strategic information. Such footage includes the new Nahang submarine, first shown on March 8 on IRINN; new uranium enrichment facilities, shown on April 11 on Channel 1, the debut of the Mesbah satellite, shown on July 20, 2005, on Channel 2, and Iranian divers planting mines, shown on April 2 on IRINN.

Channel 4 first showed the Iranian Revolutionary Guard testing its new, mass-produced flying boat, capable of "launching different types of weapons," on April 4, while Channel 2 first unveiled the Ghadir, the new Iranian submarine, on May 10.

Iran's state-controlled TV also broadcasts interviews with government and military officials that contain useful information. For example, in an interview on IRINN on April 12, the deputy commander of the Revolutionary Guards, General Ali Fadaui, announced the testing of the world's fastest underwater missile, which can "be used against fleets as well as large submarines."

A former Iranian defense minister, Admiral Ali Al-Shamkhani, appeared on Channel 2 on July 23, 2005, and gave out information about his country's space program: "The Defense Ministry has entered the fourth dimension of defense.

Previously, we operated in air, land, and sea. ... Significant steps have been made in the fields of reception stations, control, launching, launch pads, and satellite tracking. ... This is a launching pad for the Defense Ministry in order to meet the needs of the armed forces in space."

Admiral Shamkhani also appeared on Channel 2 on March 21, talking about the successful test of the Shabab 3 missile, which, he said, can be "produced like candy." And the head of the Iranian Nuclear Energy Organization, Gholam-Reza Aghazedeh, appeared on IRINN on April 13, disclosing details never before made public about Iran's nuclear project, including the creation of new nuclear facilities, centrifuges, and heavy-water manufacturing.

By monitoring Iranian TV, vital information not only about weapon components, but also about the Iranian government's nuclear negotiating strategy can be understood. When the Iranian intelligence minister, Ali Yunesi, spoke to Channel 2 on July 16, 2005, he let it slip that his government uses double agents to provide faulty information to America to "keep them busy and fool them."

Similarly, a member of the Iranian Supreme Council for National Security and the chief negotiator on nuclear affairs, Hosein Musavian, appeared on Channel 2 on August 4, 2005, and admitted that Iran used negotiations with Europe to buy time to allow for the completion of multiple components of its nuclear program.

While Iran has always insisted its desire for nuclear technology is not for military purposes, Foreign Minister Manouchehr Mottaki appeared on Channel 2 on June 7 and let out this Freudian slip: "We emphasize the peaceful nature of our nuclear weapons ... I mean ... of our nuclear power plant, and nuclear energy and activities. We have no need for nuclear weapons, as we've often said."

By listening to and watching Iran state-controlled TV, the Iranian government's true military ambitions can be understood. On April 17, the chief of the Iranian army's joint staff, General Abdorrahim Musavi, spoke on Channel 2 and discussed the possibility of a "battle with the enemy ... with America."

General Musavi spoke of Iran's "military planning," which included preparing for certain "scenarios," stressing that "we want to put America in its place" and that "America's military power is greatly overestimated."

Ignoring such statements could prove to be dangerous.

Saudi Billionaire Prince's Hate TV

June 21, 2006

The world's fifth-richest man, billionaire Saudi Prince Al-Waleed bin Talal Abd Al-Aziz, launched the new Arabic Islamic channel Al-Resalah on March 1.

The Saudi Gazette reported on March 7 that Prince Al-Waleed said his new channel was family-oriented and created to "counteract the misconceptions about Islam" and to project it as a religion of tolerance. The channel's advisory committee is made up of leading Saudi religious figures from the Senior Ulema Commission, Muslim World Congress, and the prince's Kingdom Holding Company, among others.

As he worked on launching the channel last year, Prince Al-Waleed made headlines for giving tens of millions of dollars to Harvard and Georgetown universities; he also reportedly approached other East Coast universities about making donations. The prince's Kingdom Holding Company paid for advertisements on major TV networks, including CNN, and in the New York Times and Washington Post lauding the donations and the prince's "continuous efforts to serve ... religious understanding."

"Prince Al-Waleed, Avid Supporter of Peace" was the headline of a June article in the Middle East Times International that reported on his generosity, powerful supporters, and investments in America. The article noted the prince's $100,000 donation to New York's Institute for Middle East Peace and Development and a visit to the prince's Riyadh office by dignitaries including the recently appointed treasury secretary, Henry Paulson.

Given the millions the prince has spent on public relations, Al-Resalah's excessive anti-Western content is somewhat astonishing. It is no different from any other hate-filled Saudi TV channel. Take, for example, Sheikh Ahmad Al-Kubeisi, who appeared on Al-Resalah on March 15 and said: "When there is no hope for peace, there is no alternative but to resort to the gun. ... The West's conflict with Islam and the Muslims is eternal, a preordained destiny that cannot be avoided until judgment day."

On March 7, the Arab News quoted Prince Al-Waleed as saying he plans to launch an English-language channel for a Western audience. As such, guests will include Muslim Americans - whom the channel aims to influence - such as Salah Sultan, who appeared May 17.

Mr. Sultan has been called one of America's most noted Muslim scholars. He is a signatory to the Council on American-Islamic Relations' fatwa against terrorism and is active in many American-Muslim organizations. In his appearance, Mr. Sultan praised a Yemeni sheikh, Abd al-Majid Al-Zindani, whom the American government has categorized as a specially designated global terrorist for his loyalty to Osama bin Laden and his support of Al-Qaeda. Mr. Sultan said of the attacks of September 11, 2001: "It was planned within the U.S., in order to enable the U.S. to control and terrorize the world."

Last year, Prince Al-Waleed also gave $20 million to France's leading museum, the Louvre, to establish a department of Islamic art. He told Agence France-Presse that the donation "will reinforce understanding between Western and Islamic cultures and civilizations."

Yet much of the content on his TV channel is overtly anti-Western. On March 31, the secretary-general of Al-Resalah, Sheikh Tareq Al-Suweidan, gave a speech at Dialogue between Europe and Muslims, a convention in Copenhagen that the channel was covering. "The West have done strategic mistakes ... they underestimate the power of Islam," he said. Sheikh Suweidan praised the election of Hamas and Egypt's Muslim Brotherhood, warning: "The West have no chance but to deal with Islam, and we are extending our hands in peace and dialogue - you have slapped it. We do not accept insults."

On April 6, on a program on Al-Resalah produced by the Saudi Ministry of Religious Endowments, Daw'a, and Guidance, the Saudi author Muhammad Al-Arifi said the West is inhabited by "infidels" committing adultery. "If he [the infidel] feels like having any type of sexual relations, he does ... they have organizations for homosexuals, organizations for people who marry animals," he said. Mr. Arifi claimed that Dutch women are so promiscuous that most of them are not sure who fathered their children.

In "Alwaleed: Businessman, Billionaire, Prince," a 2005 biography of the prince, Riz Khan writes that Al-Waleed "has nothing against Jews. He had always been very open about doing high-level deals with heads of key companies, even though they were Jewish." Yet Al-Resalah frequently airs anti-Semitic content such as an April 14 interview with Sheikh Hazen Sallah Abu Ismail, in which he said, "The Jews are determined to own the media in order to control the ears and eyes of the youth." Sheikh Ismail cited "statistics by the U.N." to support his claim that "82% of all attempts to corrupt humanity originate from the Jews."

Prince Al-Waleed's biography details his hard work ethic, calling him "tireless" and a person who pays "attention to the minutest detail in every transaction." If this is true, he must be aware of the anti-Western content on Al-Resalah.

'Jihadi Porn' Puts Beheadings Online

June 28, 2006

> *"The slaughter of that American [Nicholas Berg] under the slogan 'Allah Akhbar.' You know, in schools today ... 6-and 7-year-old children are shown how to slaughter, and the children have to shout 'Allah Akhbar.' There are thousands of terrified children ... who cannot sleep at night. ... Can anything good come out of this generation? Murder has become our folklore."*
>
> — HISHAM AL-IQABI, AL MUSTAKILLAH TV
> JUNE 9, 2004

Since the war on terror began, there has been a proliferation of jihadi and Islamist Web sites. In addition to sharing messages, as well as the latest speeches, sermons, and fatwas by Al-Qaeda and Islamist religious figures, the sites contain what could be best described as "Jihadi porn."

Jihadi porn includes footage of the brutal killings of Westerners taken hostage by Islamist terrorists in places such as Iraq, Saudi Arabia, Chechnya, and Pakistan. These hostages are often beheaded or shot in the back after being tortured. The Merriam-Webster dictionary describes this "pornography of violence" as "the depiction of acts in a sensational manner so as to arouse a quick, intense emotional reaction."

A new era for Islamist terrorists was heralded in January 2002 when a member of Al-Qaeda took a knife and tore off the head of a Wall Street Journal writer, Daniel Pearl, in Karachi, Pakistan. This horrific episode was filmed and put on the Internet, and countless Web sites posted it soon after.

Nicholas Berg, a 26-year-old American who went to Iraq to help build the country's telecommunications infrastructure, was beheaded by Abu Musab al-Zarqawi in May 2004. His gruesome murder also was filmed and posted on many Islamist Web sites, and the video soon became popular jihadi porn for the demented individuals who frequent such sites.

Paul Johnson Jr., who worked for Lockheed Martin in Saudi Arabia, was next. His beheading was posted on the Internet, as were images showing his severed head being held in one of his murderers' hands. After the images were released, the Al-Qaeda magazine Sawt al-Jihad published an article, titled "A Letter to the Wife of the Slain Paul Johnson From the Wife of One of the Martyrs," that warned: "We are just getting started, and the corpse of your husband shall be followed by mountains of corpses of his countrymen."

Other Americans whose beheadings were filmed and posted online include an engineer from Georgia, Jack Hensley, shown in a September 2004 video; a Michigan contractor, Eugene Armstrong, shown in a September 2004 video, and an Illinois contractor, Robert Jacob, whose beheading in Riyadh appeared in a June 2004 video.

Last week, two American soldiers became the latest American beheading victims. According to the Iraqi Defense Ministry, Private First Class Thomas Tucker and Private First Class Kristian Menchaca were reportedly "killed in a barbaric way," "slaughtered," and tortured to death, and their bodies were so mutilated that DNA tests are being performed to help identify their remains. So far no video of the killings has appeared.

Footage of the killing this past Sunday of four Russian diplomats kidnapped in Iraq, however, has appeared on the Internet. Other Westerners and Asians whose beheadings became jihadi porn include Ken Bigley, a British civil engineer who was shown in an October 2004 video; an Italian photographer, Salvatore Santoro, shown in a December 2004 video; a Japanese contractor, Akihiko Saito, shown in a May 2005 video, and a Korean translator, Kim Sun-il, shown in a December 2004 video.

Muslims also have been beheaded by their co-religionists, especially in Iraq. Notable examples include Egyptian and Algerian envoys in July 2005 videos; a Lebanese businessman, Badri Ghazi Abu Hamzah, in a January 2005 video; Turkish truck drivers and contractors in October 2004, and many other Iraqis.

The widely popular Iraqi TV terrorist confession series on Al-Iraqiya and Al-Fayhaa TV often focuses on beheadings. Notable examples include a February 23, 2005, episode featuring a captured Iraqi terrorist, Shihab Al-Sab'awi, who detailed how Syrian intelligence officials in Syria taught him to murder and behead; a March 30, 2005, episode with a captured Iraqi terrorist, Omar Allawi, who admitted he was paid $200 to film beheadings, and a April 21, 2005, episode with a captured Iraqi terrorist, Adnan Elias, who confessed to slaughtering hostages by pulling "the guts" out of their stomachs.

Those who do the beheading, and those who watch them online, share the same beliefs as the September 11 hijackers and those who send suicide bombers to attack cafes, churches, buses, and schools. As the war on terror continues, it should be remembered whom we are up against.

Murderous Mufti Slips In and Out of America

July 5, 2006

A Palestinian Authority official, Sheikh Ikrima Sabri - known over the past 12 years as the mufti of the Al-Aqsa Mosque, Islam's third-holiest shrine - was quietly removed from his position last week. The reason: ongoing provocative Friday sermons. His comments in support of Hamas and criticism of Mahmoud Abbas were enough to have him finally ousted.

Sheikh Sabri's sermons can be described as anti-Semitic and anti-Western. Therefore, it was somewhat surprising that he came to America on a speaking tour in the first week of March, visiting major American-Muslim population centers in New York, New Jersey, Virginia, Texas, Illinois, Louisiana, and Colorado.

The Denver Post of March 5 described Sheikh Sabri as "one of the most prominent Muslim leaders in the world." It was reported that his speaking tour was devoted to subjects on which he is known to have extreme views: America, the history of Jerusalem, and relations with Jews.

During an event sponsored by the Colorado Muslim Society on March 4, one member of that community was quoted by the Rocky Mountain News as saying the mufti's visit to America to "reach out to the American public" is like the pope visiting the Catholic community.

But the mufti has always been vehemently anti-American. Just two weeks before the attacks of September 11, 2001, he ended his weekly sermon on Palestinian Authority Radio by calling for "Allah [to] destroy the U.S."

Weeks after September 11, Sheikh Sabri stated that Muslims were not involved in the attacks. His September 28, 2001, sermon was devoted to "the innocence of Muslims until he is proven guilty." Instead, he pointed to possible "elements inside America who carried out these [attacks]."

Speaking about the 1993 attack on the World Trade Center during a May 25, 2001, appearance on PA TV, Sheikh Sabri gave his support to those jailed for the attack, the blind Sheikh Omar Abdel-Rahman and Ramzi Yousef, who were described as being "wrongly and aggressively jailed in America" and "charged with false claims." From the Al-Aqsa Mosque, Sheikh Sabri declared, "[We] demand the release of these prisoners, who are monotheistic Muslims."

The mufti's views about relations between Jews and Muslims can be understood by what he told Saudi Al-Majd TV on February 20, 2005: "Anyone who studies 'The Protocols of the Elders of Zion' and specifically the Talmud will discover that one of the goals of these protocols is to cause confusion in the world and to undermine security throughout the world."

Sheikh Sabri has a long history of denying any Jewish connection to Jerusalem. In an interview with Kul Al-Arab on August 18, 2000, he said, "No stone of the Al-Buraq [Wailing/Western] Wall has any relation to Judaism." On May 9, 2004, on Dream TV 2, he added: "I am the mufti of Palestine from the [Mediterranean] Sea to the [Jordan] River. ... The Al-Buraq Wall is part of Al-Aqsa Mosque. ... The Jews pray in front of the wall ... but the wall belongs to the Muslims."

Sheikh Sabri also has gone on record countless times supporting suicide attacks and inciting Palestinian Arabs to perform them. On May 25, 2001, he addressed the Jewish residents of Israel: "We tell them - inasmuch as you love life - the Muslim loves death and martyrdom. There is a great difference between he who loves the hereafter and he who loves this world. The Muslim loves death and [strives for] martyrdom. The blessed and sacred soil of Palestine has vomited all the invaders ... throughout history and it will soon vomit, with Allah's help, the [present] occupiers."

Later that week on a Tel Aviv beach, 17 Israeli teenagers were killed in a suicide bombing at the Dolphinarium.

It is somewhat ironic that the mufti would want to reach out to the West when he has so much disdain for it. One notable example includes an interview with Tanweer TV on May 5, 2005, when he criticized "the laws of the West" for punishing men "as rapists merely because he slept with his wife against her will."

Sheikh Sabri's visit to America raises many disturbing questions, among them: Who funded and initiated the trip? Why was this preacher of hate allowed to enter the country when he has called for its destruction? Did he incite his American followers behind closed doors?

To view translated segments of Palestinian Arab sermons, visit www.memritv.org.

Muslim Religious Figures Must Challenge Al-Qaeda

July 12, 2006

For the West to win the war of ideas, Muslim religious figures must openly challenge the ideology of Islamism and Al-Qaeda.

One such figure is Sheikh Abd Al-Muhsin Al-Abikan, a member of Saudi Arabia's Shura Council and consultant for the Saudi Ministry of Justice. He is well-known for publicly battling against militant Islamist ideologies and for speaking out against Al-Qaeda, jihad in Iraq, and suicide attacks.

On the one-year anniversary of the London bombings of July 7, 2005, Sheikh Abikan was quoted in the London Arabic daily Al-Sharq Al-Awsat denouncing the attacks and calling those responsible "apostates" who committed "treason."

He appeared on Saudi Channel 1 on July 22, 2005, and explained to the audience why Al-Qaeda's ideology and actions are against Islamic law: "Terrorism

as defined today in the world is prohibited by the Shariah.This terrorism seeks to cast fear upon peaceful people, Muslim and non-Muslim, whom it is forbidden to kill according to the Shariah. ... The distressing pictures and stories we see ... of beheadings ... arouse disgust."

On many occasions, Sheikh Abikan has expressed a willingness to publicly confront Osama bin Laden and his cronies. He told Al-Sharq Al-Awsat on May 24, 2005: "I say to those who believe in Al-Qaeda and its ideology of Takfir that they are being superficial. … I have challenged such people. … I've even sat and debated with such sympathizers at my home and in the mosque. … I am even ready to debate with bin Laden himself!"

Sheikh Abikan has a weekly religious program on the pan-Arab MBC channel. On April 30, 2004, he was asked about Mr. bin Laden, whom he had met years earlier. He replied, "Bin Laden [and his followers] must first turn to Allah in repentance for what they have engaged the Muslims in and for the civil strife, disasters, and tragedies caused by the instructions we hear from them through the media."

On the death of Abu Musab al-Zarqawi, he said: "The current sympathy shown by some Muslims to al-Zarqawi ... is regrettable. Al-Zarqawi, who adopted Al-Qaeda's deviant ideology, only brought evil to Muslims. Those sympathizing with the barbaric and criminal acts committed by al-Zarqawi inside Iraq and elsewhere ... chose to forget his heinous record."

"What is happening in Iraq now is not jihad," Sheikh Abikan said on MBC on July 31, 2004. "Jihad does not mean harming Muslims who cooperated with non-Muslims, as claimed. ... The problem with people is that they hear some muftis [and religious scholars] who quote from the jurisprudents things that are not relevant ... to the current situation in Iraq."

These muftis and religious scholars are dissenters who should stand trial for the damage they have caused to Iraq because of the fatwas they have issued, the sheikh told the Saudi daily Al-Riyadh on May 27, 2005. "Can we implement in Iraq what the clerics have said?" he asked on his weekly show on August 28, 2004. "Certainly not. According to the clerics, people should adhere to their imam and wage jihad against an enemy who wants to take their property and kill them. Iraq, on the other hand, is under the control of Iraqis, no matter who they are."

In the May 2005 interview with Al-Riyadh, Sheikh Abikan said the fighters blowing themselves up with explosive belts or car bombs are carrying out suicide operations, not martyrdom operations. He also clarified that killing during these operations was not an act of war but an act of treachery.

At a counterterrorism conference of leading Islamic scholars in the Egyptian resort of Sharm el-Sheikh on August 22, 2005, Sheikh Abikan corrected the religious figures in attendance on their stance vis-a-vis martyrdom: "The suicide operations that are called 'martyrdom operations' are forbidden by Islamic law. Those who carry them out, committing suicide, cannot be called martyrs, and their actions cannot be called martyrdom."

A former Saudi communications minister, Ali Bin Talal Al-Jihni, told Al-Hayat newspaper on June 7, 2005: "The Saudi religious scholar Al-Abikan is practically the only one among the Muslim scholars who raises his voice ... to say that suicide and the killing of others in Iraq must not be defined as martyrdom. Al-Abikan is not the only one who knows that [such] suicide is a crime. ... What prevents other scholars from raising their voices to support the honorable sheikh? ...Perhaps they lack his courage."

Though Sheikh Abikan's views on women's rights and reform within Islam could be considered pragmatic and not politically correct by Western standards, the West by all means should support his fight against Al-Qaeda and radical Islamism.

Arab World Divided Over Hezbollah

July 19, 2006

> *"Hezbollah is not waging the battle of Hezbollah or of Lebanon. We are waging the battle of the [Islamic] nation, whether we like it or not, whether the Lebanese like or not."*
>
> — HEZBOLLAH LEADER SHEIKH HASSAN NASRALLAH, AL-MANAR TV, JULY 16

In the aftermath of Hezbollah's attack on Israel, it appears Arab regimes are dividing into two camps. On one side are those standing with Hezbollah and Iran, such as Syria and Yemen (and the Lebanese ambassador to America, who was removed from his position last week), and on the other side is almost everyone else. The latter group clearly recognizes that chaos could engulf the entire region as Hezbollah attempts to push Lebanon to commit "martyrdom."

The Arab League has called Hezbollah's attacks on Israel "unexpected, inappropriate, and irresponsible acts," in the words of Saudi Arabia's foreign minister, Prince Saud Al-Faisal. Also, President Mubarak of Egypt and King Abdullah II of Jordan, without directly naming Hezbollah, warned last week of "the region being dragged into 'adventurism' that does not serve Arab interests."

Saudi Arabia also issued a statement on July 13 that was critical of Hezbollah's actions: "The Kingdom stood firmly with the resistance in Lebanon until Israel's occupation of Lebanon ended. ... A difference should be drawn between legitimate resistance and rash adventures carried out by elements inside the state and those behind them without consultation or coordination with Arab countries. ... The Kingdom views that it is time that these elements alone bear the full responsibility of these irresponsible acts and should alone shoulder the burden of ending the crisis they have created."

A prominent Lebanese Druze member of parliament, Walid Jumblatt, also criticized Hezbollah in an interview with the French daily Le Figaro on July 13: "Hezbollah played a very dangerous game by kidnapping these two soldiers. ... Israel has withdrawn from Lebanon. These kidnappings took place outside our country's borders. Hezbollah is raising the stakes, with the support of Iran and Syria, which in the past year has become a satellite state of Tehran. ... Hezbollah will have to explain itself to the Lebanese. ... [We need to decide] who makes the decisions regarding war and peace in this country."

The Arab press also has been highly critical of Hezbollah. The editor in chief of the Kuwaiti daily Al-Seyassah, Ahmed Jarallah, has written a number of editorials on the subject. On July 15, he wrote that Israel's operations against Hezbollah are in the interest of the world: "People of Arab countries, especially the Lebanese and Palestinians, have been held hostage for a long time in the name of 'resisting Israel.' ... Unfortunately, we must admit that in such a war the only way to get rid of 'these irregular phenomena' is what Israel is doing. The operations of Israel in ...Lebanon are in the interest of people of Arab countries and the international community."

In an editorial the next day, Mr. Jarallah wrote: "Nasrallah has done some grave miscalculations under instructions from the outside. Nasrallah has dragged Lebanon and its people into misfortune. In spite of the destruction caused by Israel, Lebanese politicians don't want to be frank with their people and tell them that they should not support Nasrallah's decision to declare war on Israel. Nasrallah has hijacked the authority of the Lebanese government to have control over the people of Lebanon while Lebanese politicians continue to remain mute spectators without voicing their true feelings. ... The fate of the Lebanese is in the hands of a handful of reckless adventurers, who have prevented Arabs from making well-judged decisions. ... Iran and Syria are fighting the international community, especially the United States, in Lebanon and Palestine. Nobody is benefiting from this conflict, except Tehran and Damascus."

Many other members of the Arab press have been speaking out against Hezbollah. They all seem to recognize that Hezbollah's actions could potentially destroy Lebanon and cause an enormous amount of damage to the entire region.

Hezbollah's Nazi Tactics

July 26, 2006

"Just like Hitler fought the Jews, we are a great Islamic nation of jihad, and we too should fight the Jews and burn them."

— HISHAM SHAMAS, POLITICAL SCIENCE STUDENT, AT A SYMPOSIUM HOSTED BY HEZBOLLAH'S AL-MANAR TV AT LEBANON'S LARGEST AND ONLY GOVERNMENT-RUN UNIVERSITY, UNIVERSITÉ LIBANAISE, NOVEMBER 29, 2005

In June 1944, a Red Cross delegation was invited by Nazi Germany to Theresienstadt, a "spa town" where elderly German Jews could "retire" in safety, according to Nazi propaganda. However, it was really a transit camp for Jews on their way to Auschwitz.

In preparation for the delegation's visit, the camp underwent a beautification program, with gardens planted and barracks renovated. The Nazis produced a film using Jewish prisoners to show the world how well they were being treated (the entire "cast" would later be sent to Auschwitz). Theresienstadt served an important propaganda function for the Germans, and it is known by historians as an "an elaborate hoax."

Taking a page directly from the Nazis' propaganda playbook some 62 years later, the terror organization Hezbollah is manipulating the press in the same way. Hezbollah's press officer, Hussein Nabulsi, has been giving Western press and broadcast outlets, including CNN, the Boston Globe, and the Chicago Tribune, daily tours of bombed-out Hezbollah-controlled southern Beirut.

During the July 18 broadcast of Wolf Blitzer's "Situation Room" on CNN, Mr. Nabulsi took a CNN correspondent, Nic Robertson, to a bombed-out area he said was "only inhabited by innocent civilians." Mr. Robertson was overtly sympathetic to Hezbollah's claims, telling viewers, "We didn't see any evidence of military infrastructure."

Mr. Roberston's sympathy for Hezbollah later was criticized on multiple blogs; even the host of CNN's "Reliable Sources," Howard Kurtz, asked his colleague, "To what extent do you feel like you're being used to put up the pictures they [Hezbollah] want?"

In a report for the July 21 Boston Globe, Thanassis Cambanis wrote about his tour of the same area with Mr. Nabulsi, who admitted that his office — Hezbollah's press office — which he said was located in a "civilian area," was bombed. The Globe reporter wrote that the area "is full of obvious Hezbollah targets, including the organization's headquarters, its media office ... nestled among thousands of civilian homes."

Just as the Nazis had their weekly newspaper, Der Stürmer — which was a significant part of their propaganda apparatus, spreading venomous anti-Semitic caricatures and themes, including the blood libel and the "Protocols of the Elders of Zion" — Hezbollah has its own radio station and TV channel, Al-Manar.

Al-Manar is similarly a vehicle for spreading classic anti-Semitic subject matter, from Nazi forgeries to the blood libel and attacks on Judaism. The channel also frequently airs programs devoted to the killing of Jews and destruction of Israel. (See www.memritv.org to view examples.)

Hezbollah celebrates Holocaust denial, as well. "Jews invented the legend of the Holocaust," the leader of Hezbollah, Sheikh Hassan Nasrallah, said on April 9, 2000. During an appearance on Al-Manar on February 3, Sheikh Nasrallah called Europe's leading Holocaust denier, Roger Garaudy, "a great French philosopher." On February 23, Sheikh Nasrallah appeared on Al-Manar and praised another leading European Holocaust denier, David Irving, for having "denied the existence of gas chambers."

While Hezbollah has embraced much of Nazi Germany's anti-Semitism, it is not the heir apparent of the Third Reich. That distinction belongs to the Islamic Republic of Iran, which today has set out to launch World War III.

Just as Nazi Germany used its cohorts in France and Italy to carry out its killing, Hezbollah is following orders from its master Iran, which created it on February 16, 1985. Hezbollah's official manifesto includes a picture of "our leader," Ayatollah Khomeini, on its back cover. A Washington Post article from that year reported that Iranian Revolutionary Guards were "carrying out their missionary work, indoctrinating the Lebanese Shiites in the spiritual and political teachings of Khomeini."

As the Iranian parliament speaker, Gholam-Ali Haddad Adel, told the Iranian News Channel on July 18, "To Nasrallah we say ... the blood of Imam Khomeini rages in his veins."

Sheikh Nasrallah's deputy, Sheikh Naim Qasem, quoted the Iranian regime in explaining his organization's position on Israel and jihad in a June 2, 2002, interview on Al-Manar: "Muslims should annihilate Israel. ... Imam Khomeini said, 'The goal of this virus [Israel] that was planted in the heart of the Islamic world ... The solution is in annihilating the virus.'"

Anyone who has witnessed Iranian crowds chanting "Death to America" and "Death to Israel" during religious rallies led by the country's ayatollahs, or at Lebanese rallies led by Sheikh Nasrallah, can see how similar they are to the rallies Nazi Germany held — only then, the chants were "Heil Hitler."

An Islamic Republic Is Hezbollah's Aim

August 2, 2006

"An Open Letter: The Hezbollah Program," the terrorist organization's manifesto, was published by the Lebanese daily Al-Safir on February 16, 1985. It emphasized "the necessity for the destruction of Israel," as well as Hezbollah's other main objective, an Islamic regime in Lebanon with laws based on Shariah and with non-Muslims called upon to "embrace Islam."

Some Middle East analysts claim that Hezbollah gave up its aim of creating an Iranian-style Islamic republic in Lebanon years ago. They are mistaken. Since its inception, Hezbollah has always sought to further an Islamic way of life for the people of Lebanon. This began after the group was founded, when its leaders imposed strict codes of Islamic behavior on areas in the south of the country.

Whereas Hezbollah stayed out of politics in its early days, it has since become a political force, with 14 delegates in the 128-seat Lebanese national assembly. Hezbollah has likely decided to follow in the footsteps of Hamas and could now, with its swelling popularity, be seeking to take over the Lebanese government. If this succeeds, it would not be surprising if one day Hezbollah attempts to impose its religious program on the entire country.

A columnist for the London Arabic daily Al-Hayat, Raghida Dergham, wrote on July 28 about possible scenarios that Sheikh Hassan Nasrallah is planning. "Toppling the Lebanese government may be the common desire of Hezbollah," she wrote. "If this is true, the movement is fighting a war to topple the state and turn Lebanon into a domain of the resistance, whether the Lebanese like it or not. Assuming that this is the case, Hezbollah has chosen to end the dialogue, defy resolution 1559 to quash it, and spread chaos and destruction in Lebanon."

"Whether the Lebanese like it or not" is what Sheikh Nasrallah said in a speech aired on Al-Manar TV on July 16. "Hezbollah is not waging the battle of Hezbollah or of Lebanon. We are waging the battle of the nation, whether we like it or not, whether the Lebanese like it or not. Lebanon and the resistance of Lebanon are waging the battle of the nation."

President Ahmadinejad of Iran likewise explained the importance of Hezbollah's battle in an interview that aired on the Iranian News Channel on July 23. "In my opinion, Lebanon is the scene of a historic test that will determine the future of humanity," he said. Hassan Hanizadeh wrote in the Tehran Times on the same day,"The outcome of this war will definitely determine the region's fate," and Iran's supreme leader, Ayatollah Khamenei, explained what is at stake in a July 17 speech."Lebanon, a country that was supposed to become the main base of Western culture in the Arab world, has turned into a base for jihadi and resistance cultures," he boasted.

Many in the Arab world, however, including secular Muslims and non-Muslims who have begun to support Hezbollah, may not understand this point. As many Western press outlets have been reporting, the war between Israel and Hezbollah has created widespread public support for the terrorist group. Lebanon's Prime Minister Siniora even thanked Hezbollah on Sunday for its "sacrifices"in its war against Israel: "We are in a strong position, and I thank the Sayyed for his efforts," Mr. Siniora said.

Hezbollah's increased popularity was the subject of a discussion on CNN by a Sarah Lawrence College professor, Fawaz Gerges, on July 30: "Hezbollah has become now a symbol of resistance and defiance, not just for Lebanese, but I think for many Arabs and Muslims. And the leader of Hezbollah, Hassan Nasrallah, I think has become one of the most undisputed Arab heroes for Arabs and Muslims."

Mr. Gerges added, "Just this week ... a very prestigious Beirut research institute, along with the Lebanese-American University, did a survey of thousands of Lebanese.And the findings, they show almost 87% of all Lebanese, Shiites, Sunnis, and Christians support Hezbollah. ... This tells you that there has been a swell of support for Hezbollah inside Lebanon. ... The longer the war continues, the stronger Hezbollah would become, and, in fact, the weaker the democratically elected Lebanese government would be."

Hezbollah supporters should be careful. After all, the alliance of urban, middle-class, liberal, leftist, secular, and religious masses supported the Iranian Revolution in 1979 and thought Ayatollah Khomeini would bring fundamental rights and freedom to their country.

To this day, they are paying for that mistake.

Iran and Hezbollah Want To Force Jews To Exit Israel

August 9, 2006

> *"All those Jews who came here [to Israel] from all over the world must return to the places from where they came. The Jews from Germany should return to Germany."*
>
> — IKRIMA SABRI, A FORMER PALESTINIAN AUTHORITY MUFTI, DIE WELT, JANUARY 2001

Jews should leave Israel and "return" to "where they come from." This is one of the aims of Hezbollah and Iran's war against Israel. The strategy to destroy the Jewish state is based on a plan in which Hezbollah, with weapons provided by Iran via Syria, will terrorize Israelis in hopes that it will flee. Both Iran and Hezbollah openly state that this is their ultimate goal.

Over the past year, Arab and Muslim leaders have suggested Alaska, Canada, and anywhere in Europe as alternative territories for the Jewish state. A former Malaysian prime minister, Mahathir Mohammad, appeared on New TV on February 15 and called for the Jews to leave Israel, suggesting, "How about giving a little bit of Texas to make the state of Israel?"

The leader of Hezbollah, Sheikh Hassan Nasrallah, gave an interview to Al-Manar TV on February 18 in which he was asked, "Is there no end to this hostility" with Israel? He answered: "Yes, there is an end. If the Zionists leave our lands [i.e. Israel] ... this conflict will come to an end."

Sheikh Nasrallah elaborated on this concept in a May 23 speech on Al-Manar. He explained that Israel's most important weaknesses are "the fact that it is not deeply rooted" and "its society is not homogeneous." He said the Jews "are bound together by a baseless and unfounded myth,"

"Another weakness of this entity is that its people came because they were promised security, peace, and a life in the land of milk and honey. But if they encounter something else, they will leave this land," he said.

Hezbollah's representative in Iran, Hossein Safiadeen, told Islamic Republic of

Iran Broadcasting on July 24: "The people who came to Israel, [they] moved there to live, not to die. If we continue to attack, they will leave."

Over the past year, in addition to President Ahmadinejad's calls for Israel to be wiped off the map, the Iranian leader also has been quoted calling on Jews to leave the country:

- "The Europeans can give parts of their territories to Israel. ... Germany and Austria in particular could place some of their provinces at Israelis' disposal." Al-Alam TV, December 8
- "If you [Europeans] are correct in saying that you killed 6 million Jews in World War II ... it is only appropriate that you place a piece of your land at their disposal — a piece of Europe, of America, of Canada, or of Alaska — so they can establish their own state." IRINN TV, December 14
- "Why do you think they [the Jews] are safe in Palestine? These people, who left their homelands as a result of your pressure and anti-Judaism, went to a country that did not belong to them. Different minorities from different countries came to live side by side, just like pieces of paper joined together with a paper clip. ... Open the gates of this big prison. Allow these people to decide freely, and you will see that they will return to their homelands." IRINN TV, April 24. The Iranian parliament speaker, Gholam-Ali Haddad Adel, gave an interview to IRINN TV on July 18 in which he also explained, "Following World War II, they established an artificial, false, and fictitious state called Israel ... claiming this was support for the Jews of the world. ... They came to Palestine. ... Today is the day you will flee occupied Palestine. You must return to your homes."

Of late, propaganda from Iranian press and broadcast outlets has made numerous claims that Jews are leaving Israel. "The reverse immigration of Jews to Europe has begun," Iran Daily reported on July 27.

In an August 7 article in the Tehran Times, titled "No More Milk and Honey in Israel," Hassan Hanizadeh wrote: "The pace of emigration from the occupied territories to Europe and North America has significantly accelerated. So, the old saying that Israel is that land of milk and honey ... is now relegated to the dustbin of history."

It should be noted that during the second intifada, the Arab and Iranian press outlets also claimed that the Jews were leaving Israel en masse. In fact, the Jewish population increased during this period.

Islamic Leaders Urge Children To Be Bombers

August 16, 2006

> *"Young Israelis dream of being inventors, and their role models are the Israeli innovators who made it to the Nasdaq. Hezbollah youth dream of being martyrs, and their role models are Islamic militants who made it to the Next World."*
>
> — THOMAS FRIEDMAN, NEW YORK TIMES, AUGUST 9

The act of infanticide, the practice of intentionally causing the death of an infant, is rarely practiced today. Yet in the Middle East, it has taken on a new form. Under the direction of some leading Muslim religious figures, some parents in the region are encouraging their children to commit suicide as a religious act — then celebrating it.

Since Sheikh Hassan Nasrallah has led Hezbollah, the terrorist organization has put the death of Lebanese children on a pedestal. His Ashura speech of May 1997 praised parents who beseeched Allah to "bless" their children with the honor of martyrdom.

Four months later, Sheikh Nasrallah's 18 year-old son became a "martyr" after battling Israeli forces. The Hezbollah leader appeared on Al-Manar TV and thanked Allah for his son's "martyrdom," saying it brought him "the greatest feeling of joy that a father can know."

Many Lebanese Shiites were inspired by his words and to this day they want the same for their children. Sanaa Younes told the San Jose Mercury News on August 3, "He gave his son to Islam. ... It's what every parent would want."

On July 23, another Shiite mother told the Boston Globe, "We are ready to give our children for Sheikh Nasrallah."

Hezbollah begins brainwashing young children to be willing to kill themselves from an early age. "Hezbollah," a 1997 book on the terrorist group by a Reuters Middle East correspondent, Hala Jaber, details how the terror organization and its TV channel incite Muslim youth to terrorism: "Al-Manar ... is dominated by religious programs. Pictures and names of martyrs are screened, supported by verses from the Koran which glorify such deaths. The aim is simple: to indoctrinate the minds of the young ... with the idea that those who seek martyrdom will be rewarded with more pleasure than can ever be achieved during this earthly lifetime."

In an interview with the Boston Globe on August 1, two young members of Hezbollah, "Hussein" and "Hamid," described their meticulous training, which began at age 14. Hussein explained, "It is the matter that we are not afraid of death. ...This is the center of the training of a fighter, to make him unafraid of death, so you prefer to die rather than live humiliated."

The mother of a Hezbollah martyr, Bassel Al-Din, appeared on Al-Manar on May 22, 2005. She wept with happiness as she told the channel what happened when her child became a martyr: "Bassel had a wish. ... Whenever I told him I wanted to marry him off, he would say, 'Yes, mother, you'll marry me off like this in paradise.' And indeed, the martyr Bassel got married in paradise. I congratulate the black-eyed virgins who took Bassel from me."

A "Mother's Day Special" on Al-Manar on November 11, 2004, featured comments from many mothers of martyrs. "All I want is martyrdom. I'm willing for all my children to become martyrs," one said.

"It's true I sacrificed a son, but others have sacrificed two or three. I hope more of my sons will become martyrs," another said.

As "Nasrallah, a Name for Arab Children," an article on the Islamic Republic of Iran Broadcasting News Network, reported August 13, over the past month, many parents in the Arab world have become so smitten with Sheikh Nasrallah that they are naming their babies after him.This is especially true in the Palestinian Arab territories, where more than 50 newborns have been given the name Nasrallah in the past few weeks.

Sheikh Nasrallah's Islamo-infanticide is also popular in Iran, where it originates. The speaker of the Iranian parliament, Gholam-Ali Haddad Adel, praised the "lion cubs of Hezbollah" in a speech that aired July 18 on the Iranian News Channel.

Iran's Fars News Agency covered a July 31 rally by representatives of Lebanese women's organizations who chanted, "Lebanese people are proud of being martyred." Pictures from the rally show Iranian teenage girls sporting the latest suicide belts and holding posters of their heartthrob, Sheikh Nasrallah. Some of the girls also carried pictures of female Palestinian Arab suicide bombers such as Hanadi Jaradat and Wafaa Idris.

A daughter of Ayatollah Khomeini, Dr. Zahra Mostafavi, sent a letter to Sheikh Nasrallah on July 31 that appeared in the Tehran Times. "The jihad you have commenced at present is not to defend a land alone, but the entity of Islam, Quran, and all Muslims. ... The only bitter and heartrending side of the holy jihad is the martyrdom of the Lebanese and Palestinian hero children ... whose martyrdom is moving for every free man," she wrote.

Until Muslim religious leaders in countries such as Lebanon and Iran speak out against child martyrs, this phenomenon can only be expected to grow.

Arab Press Outlets Mixed On Hezbollah 'Victory'

August 23, 2006

Qatar's emir, Sheikh Hamad bin Khalifa al-Thani, surprised many on Monday when he praised Hezbollah's actions against Israel as "the first Arab victory, which we have wanted for many years."

While the general consensus has been that Hezbollah is the undisputed victor of its recent conflict with Israel, this opinion is not entirely shared by the Arab press.

Some leading writers have said Hezbollah actually lost, and they are comparing the declaration of victory by the terrorist organization's leader, Sheikh Hassan Nasrallah, to those of Iraq's Saddam Hussein and Egypt's Gamal Nasser, who previously claimed to have defeated America and Israel.

"Soon Syria and Iran will start spreading illusions of victory... doing a repeat of 1967 when Gamal Abdul Nasser and other intellectual terrorist groups attempted to turn a disgracing defeat into a temporary setback," the editor in chief of the Kuwaiti daily Al-Siyassa, Ahmed Al-Jarallah, wrote in "Confront Reality, Accept Defeat," an article in the paper's August 13 edition. "Hezbollah's defeat has affected everything in Lebanon — including its people, buildings, and infrastructure. ... The Arab world must face reality and accept it has been defeated. ... Arabs should understand it will be a fatal mistake to challenge a powerful enemy and lead their countries into a dreadful adventure like Nasrallah and then urge the Arab and Islamic worlds to come to the rescue."

"Today, after the war has ended, Nasrallah, Ahmadinejad, and Assad have come out to announce 'a strategic victory.' Whose victory is this?" the editor in chief of Al-Sharq Al-Awsat, Tariq Alhomayed, wrote August 16 in an article titled "A Fake Celebration." "Nasrallah, who started his fight, claiming, 'We don't need anyone's support,' changed his speeches to demand a cease-fire ... after the fighting stopped, he spoke of victory, bluntly warning anyone from trying to disarm Hezbollah. ... Today, after Lebanon has been destroyed, it is proclaiming 'a strategic victory'!"

"Another fantasy" is how a Libyan intellectual, Dr. Muhammaed Al-Huni, described claims that Hezbollah won. In an article he authored for the Arabic reformist Web site Elaph.com on July 26, he too compared how the Arab world believed that Saddam would deliver a humiliating defeat to America to believing now that Hezbollah and Sheikh Nasrallah defeated Israel.

"Nasrallah's gift to the Lebanese people … devastation and destruction that Lebanon has never witnessed anything like it before," an Egyptian journalist, Subhy Fouad, wrote in "Which Victory Are They Talking About?" an August 16 article on Elaph.com."What is the historical and strategic victory achieved by Hassan Nasrallah over Israel that he gives to the Lebanese people? ... I would have wished that Nasrallah had announced his responsibility for the wrong decision in involving the Lebanese people against their will in a war against Israel knowing its consequences in advance."

Massoud Behnoud furthered this argument on August 14 in the Iranian reformist daily Rooz. "The more damage that is inflicted in Lebanon, the more epic becomes the words and reports of newspapers such as Keyhan that claims the victory of Hezbollah. Interestingly, they portray an image as if the whole world is calling for the victory of Hezbollah and that the world champion today is no other than Lebanon's seyed, Hassan Nasrallah. So far from reality," he wrote.

"Is it logical that people are saying that Hezbollah's jihad and the defeat it is handing to Israel did not liberate Lebanon, but rather set us back 50 years, and handed over at least part of southern Lebanon to French forces?" the Egyptian liberal Mamound Fandy's wrote in the August 7 edition of Al-Sharq Al-Awsat in his "ironic" take on the outcome of the conflict.

"Hezbollah's situation is not like that of Israel, which has suffered much less destruction than Lebanon has," Hazem Saghieh wrote August 14 in the London-based Arabic-language daily Al-Hayat. "Israel has not given up. ... Instead, it threatens us with wide-scale land invasion. … But we delude ourselves by talking about 'the beginning of the end of Israel,' portraying the loss as a defeat, or putting Israel's loss on par with our victory, while witnessing what is becoming of Lebanon."

Mr. Saghieh noted, "The Lebanese, who hailed the 'beginning of the end of the American Empire' after U.S. Marines withdrew from Lebanon, recall that this 'empire' succeeded, after only seven years, in defeating the Soviet Union without shooting a single bullet. How, then, can we say that Hezbollah has triumphed?"

Clearly, it is too soon to declare a winner in the recent Hezbollah-Israeli conflict. Only history books will determine this.

Arab Press Says Jews Perpetrated 9/11 Attacks

August 30, 2006

> *"Is it conceivable that Muslims did the killing on 9/11? Yes, they did. They killed 3,000 people."*
>
> — AHMAD AL-RUB'I, A FORMER KUWAITI MINISTER OF EDUCATION, AL-RAI TV, MARCH 27

Twelve days from now will mark the fifth anniversary of the attacks of September 11, 2001. Since the attacks, the Middle East Media Research Institute has documented what the Arab and Iranian press has said about that day.

Its response to the attacks was immediate. "When the towers collapsed ... my lungs filled with air. I breathed in relief, as I'd never breathed before," a missive from the chairman of the Syrian Arab Writers Association read. "Millions across the world shouted in joy: America was hit," the independent Egyptian weekly Al-Maydan wrote. "I have a sneaking suspicion that George W. Bush was involved in the operations of September 11, as was Colin Powell," a columnist for the Arabic-language daily Al-Sharq Al-Awsat, Samir Atallah, said.

Prominent journalists, members of academia, and leading religious figures shaped such conspiracy theories. Furthermore, Arab government officials were responsible in part for creating and spreading conspiracies about what "really" happened.

"I find it hard to believe that people who were learning to fly in Florida could, within a year and a half, fly large commercial airlines and hit with accuracy the towers of the World Trade Center, which would appear, to the pilot from the air, the size of a pencil," President Mubarak of Egypt told Al-Ahram Weekly on October 25, 2001. "Only a professional pilot could carry out this mission, not someone who learned to fly for 18 months in Florida."

Two weeks after the attacks, Sheikh Muhammad al-Gameia — an Egyptian imam at the Islamic Cultural Center of New York and the American representative to Al-Azhar University, the highest religious authority in Sunni Islam — gave an interview to one of the school's Web sites. He said, among other things, "If the

Americans knew that the Jews carried out the September 11 attacks, they would do to them what Hitler did."

More recently, an American Muslim scholar who is active in prominent American Muslim organizations such as the Council on American-Islamic Relations, Salah Sultan, gave an interview to Saudi Al-Resala TV. "September 11 could not have been carried out entirely from outside [America] — by Muslims or others.... The entire thing was of a large scale and was planned within America in order to enable America to control and terrorize the entire world," he said on May 17.

Notable conspiracy theories from the Arab world and Iran between 2001 and 2004 put the blame for the attacks on Jews, Zionists, Christian Zionists, born again Christians, the Vatican, Mossad, CIA, National Security Council, white supremacist groups, Britain, President Bush, Vice President Cheney, Benjamin Netanyahu, and many others.

These conspiracies have continued over the past year. "The 9/11 story is, in fact, a Zionist conspiracy," an Egyptian cleric, Abd Al-Sabur Shaheen, said on Al-Nas TV on July 10.

Writing in the Oman daily Al-Watan on June 26, Abdallah Hammouda discussed the possibility that neoconservative Christians and Zionists planned and implemented the attacks.

The "real facts are not available"to explain what really happened on September 11, but besides Al-Qaeda, it is possible Jews or even the "Americans themselves were behind it," the chairman of the Somali Islamic Courts Union, Sheikh Sherif Sheikh Ahmed, told the Somaliland Times on June 9.

"I said that 9/11 was carried out by American agents," the deputy chairman of the Egyptian Parliamentary Committee for Defense and National Security, Muhammad Abd Al-Fattah, told Ein TV on May 2.

"I am one of the those who believe these events" of September 11 "were fabricated from the outset," an Egyptian cleric who once lectured in America, Hazem Sallah Abu Ismail, said on Saudi Al-Resala TV on April 14.

Throughout September 2005, Al-Jazeera aired a special titled "The Truth Behind 9/11." Part IV, which aired September 30, was devoted to the Mossad's involvement, including "agents ... dancing and cheering in front of the World Trade Center."

On September 13, 2005, Iran's Jaame Jam 1 TV aired a program on September 11 conspiracies, including the "true" passengers on the planes that hit the World Trade Center and the Pentagon and why 4,000 Jews did not show up to work that day. An Iranian TV documentary on Sahar 1 TV on September 11, 2005, suggested that the American government attacked its own people.

A retired Egyptian general, Muhammed Khalef, said on Al-Mihwar TV on September 11, 2005, that the planning for September 11, 2001, began in 1999 at the

National Defense University. He claimed that Mr. Cheney told Mr. Bush, "It was an inside the White House job."

In November 2002, Ain Al-Yaqeen quoted the powerful Saudi interior minister, Prince Nayef, as saying: "It is impossible that 19 youths carried out the operation of September 11, or that bin Laden or Al-Qaeda did that alone. ... I think [the Zionists] are behind these events." Prince Nayef's statement was considered shocking and was immediately condemned. As NBC reported, it completely undercut a $10 million Saudi public relations campaign.

Unfortunately, as we approach the fifth anniversary of the attacks, statements such as Prince Nayef's are now readily accepted in the Middle East. On June 22, Pew released the results of a survey that asked Muslims if they thought Arabs carried out the September 11 attacks. Only 17% of British Muslims; 48% of French Muslims; 16% of Indonesians; 32% of Egyptians; 16% of Turks; 39% of Jordanians, and 15% of Pakistanis answered "yes."

Terrorists Promise More Attacks Like 9/11

September 6, 2006

Last year, to mark the fourth anniversary of the attacks of September 11, 2001, this column included translated proclamations by Al-Qaeda leaders promising the future killing of millions of Americans and the collapse of America, as well as statements by Arab writers calling the war on terror "a chapter in World War III."

Some pundits have chosen to reject such analogies. But, as Osama bin Laden wrote in his "Letter to the Muslims of Iraq and [Islamic] Nation" on December 27, 2004: "The conflict in the West is a fateful war between unbelief and Islam, between the army of Muhammad, the army of belief, and the people of the cross. ... The important, tremendous, and dangerous issue today in the entire world is this Third World War."

Much has happened in the war on terror over the past year. The Arab press has been full of articles on the end of America. Walid Nouhad wrote in the Bahrain daily Al-Wasat on June 20 on the coming "breakup of the United States ... in the same fashion as those empires that came before it." America's "destiny" is to be "just like the Soviet Union" when "it falls apart," Saud Mokhtar wrote in the Saudi daily Al-Madinah on July 10.

Other articles in the Arab press have featured interviews with religious and Islamist figures making crude calls for the West to convert to Islam.

"Our aim is to put down roots in the European continent, and to act quietly and in accordance with the laws, so that one day we may see all of Europe Muslim!" the Turkish fundamentalist leader Nijmuddin Erbakan, a former Turkish prime minister who headed the Islamic Welfare Party, said, according to an April piece on the Arabic reformist Web site Elaph.com by a Germany-based Kurdish journalist, Tarek Hamo.

Previous statements like "Islam will return to Europe as a conqueror" and take over Rome — by one of the most influential clerics in Sunni Islam, Sheikh Yousef Al-Qaradhawi — are well-known.

In another headline-making instance, Sheikh Ibrahim Mudeiris said on Palestinian Authority TV on May 13, 2005: "The day will come when we will rule America. The day will come when we will rule Britain and the entire world."

The Ayatollah Ahmad Husseini Al-Baghdadi gave a series of statements from April through May to Al-Jazeera, Syrian TV, and ANB TV, in which he said, "We [Muslims] will conquer the world, so that 'There is no God but Allah, and Muhammad is the Prophet of Allah' will be triumphant over the domes of Moscow, Washington, and Paris."

He added, "We will annihilate America."

The Internet edition of the Norwegian daily Dagbladet interviewed a Kurdish Iraqi Islamist, Mullah Krekar, on March 13. "Look at the development of the population in Europe, where the number of Muslims increases like mosquitoes. … By 2050, 30% of the European population will be Muslim," he said.

In a speech that aired April 10 on Al-Jazeera, the Libyan leader Muammar Gadhafi similarly stated, "We have 50 million Muslims in Europe. There are signs that Allah will grant Islam victory in Europe — without swords, without guns, without conquests … [they will] turn it into a Muslim continent within a few decades. ... Europe is in a predicament, and so is America. They should agree to become Islamic in the course of time, or else declare war on the Muslims."

Sheikh Omar bin Bakri, a Beirut-based, Syrian-born British Islamist and former leader of Al-Muhajirun, an organization in Britain, gave an interview to the London daily Al-Sharq Al-Awsat on July 7. "What is today called 'Londonistan' is in fact 'Heretistan,' that is, dar al-kufr [the abode of heresy]," he said. "I think that loyal Muslims in Britain will one day turn it, with Allah's help, into 'Islamistan,' that is, dar al-islam [the abode of Islam]. … Then the great Islamic dream will be fulfilled — that we will see the banner [proclaiming] 'There is no God but Allah' flying over Big Ben and the British Parliament."

Ahmad Abu Matar, a Palestinian Arab academic who lives in Oslo, criticized such statements in an article on Elaph.com on April 30, 2005. He also criticized Muslim groups in Europe, including the Islamic Liberation Party, which announced "from London its political platform — to establish the Islamic caliphate over all corners of the earth — and declares that the party will suggest to the queen of England that she convert to Islam."

This year also saw calls for other world leaders to convert to Islam. On June 15, the Indonesian militant cleric Abu Bakar Bashir called on Prime Minister Howard of Australia and President Bush to convert, saying it was "the only way to save their souls." Iran's Ayatollah Jannati called for Fidel Castro to convert this year, and President Ahmadinejad of Iran invited Mr. Bush to embrace Islam in a now infamous May letter.

Al-Qaeda leaders also continued their threats against the West and invited

Westerners to convert to Islam. Deputy leader Ayman al-Zawahiri's speech of July 27 called for all "former Muslims countries," including Spain, to be liberated. This followed an interview on the fourth anniversary of September 11 in which he said, "We invite the Americans to join Islam ... and to realize that their current faith ... will only lead them to destruction."

This past week, Mr. Zawahiri appeared on another video with Al-Qaeda's Adam Gadahn, aka Azzam the American, who stated: "The truth of Islam is sweeping across America and constitutes a fearsome challenge to ... the survival of the crusader state [America]."

As the war on terror continues, it is worth quoting a line from Mr. Bush's speech yesterday: "We know what the terrorists intend to do because they've told us — and we need to take their words seriously."

A Jewish Musician's Son Joins Al-Qaeda's Ranks

September 13, 2006

It seems inconceivable that some 28 years ago in California a little-known psychedelic Jewish musician and his wife gave birth to Adam Pearlman, who would become one of America's most notorious traitors in the war on terror. This Southern California boy has emerged as a leading instigator of a religious war between Christians and Jews on one side and the Muslim world on the other.

News outlets reported on September 8 that a sealed indictment accused Adam, now known as Al-Qaeda operative "Azzam the American," of helping the terror organization with communications and propaganda, serving as an English translator, and providing information about American culture and vulnerabilities. Treason charges are pending.

In 2004, at the age of 26, Azzam was first put on the FBI's most wanted list; his aliases included Abu Suhayb Al-Amriki, Abu Suhayb, Yihya Majadin Adams, and Yahya. He reportedly received training in Afghanistan terrorist camps, and Newsweek reported that he was sent by the mastermind of the attacks of September 11, 2001, Khalid Sheikh Muhammad, on a mission to join a suicide attack outside Baltimore.

In time for the fifth anniversary of September 11, Azzam appeared on Al-Sahab — the press and broadcast production house of Al-Qaeda and its main propaganda vehicle — attacking America as the "Zionist-Crusader-Fascist" entity. The majority of his speech was a verbal assault on Christianity, which he called a "hollow shell of a religion, whose followers cling to an empty faith, and a false conviction of their inevitable salvation."

Azzam was given the honor of being introduced by Al-Qaeda's deputy leader, Ayman al-Zawahiri, in the September 2 tape, and directed his speech "To America and the rest of Christendom." "Anyone who pays any attention to the messages of the leaders of the jihad, like Sheikh Osama bin Laden and Sheikh Ayman al-Zawahiri ... will know that they have been consistent in inviting the Americans and other unbelievers to Islam. ...We would prefer that all Americans and all the other unbelievers ... accept the truth. Islam encourages us to invite even the most obstinate enemies and give them a chance to repent," he said.

Adam Pearlman's religious journey began when his family converted to Christianity and took the last name Gadahn. He first came in contact with Islam when his father slaughtered goats and sold the meat to Muslim halal markets. As a pimply, head-banging teenager, Adam embraced Islam at an Orange County mosque when he was just 17.

Until this summer, an essay by "Adam Pearlman" was posted on the University of Southern California's Muslim Student Association compendium of Muslim texts; it has been quoted widely in the press. Titled "Becoming Muslim," it states: "As I began reading English translations of the Qur'an, I became more and more convinced of the truth and authenticity of Allah's teachings. ... Having been around Muslims in my formative years, I knew well that they were not the bloodthirsty, barbaric terrorists that the news media and the televangelists paint them to be."

In October 2004, with his face covered, Azzam appeared in his first Al-Qaeda video and declared, "Allah willing, the streets of America will run red with blood."

Azzam's next video, again with his face covered, was aired on Al-Jazeera on September 12, 2005, "Four years after the blessed raids on New York and Washington, we find the people of the West continuing to speculate about the ... historic events. ... We find them uncertain about which steps or actions they must take to achieve the restoration of the security they once enjoyed," he said.

He explained that the numerous audiotapes, videotapes, and communiqués Al-Qaeda has issued since September 11 "have been released to explain and propound the nature and goals of the worldwide jihad against America and the Crusaders and convey our legitimate demands."

Azzam was not heard from again until July 8, when counterterrorism officials said he had moved up in rank within Al-Qaeda and was now a key official behind Al-Sahab's operations. He appeared in a video for the first time without his mask, alongside footage of the London suicide bomber Shehzad Tanweer and Mr. Zawahiri, and warned: "When we bomb their cities and civilians ... no sane Muslim should shed tears for them."

Some counterterrorism analysts have downplayed Azzam's warnings and other English-language messages directed to his fellow Americans on behalf of Al-Qaeda.

However, as they include what is commonly referred to in different periods of Islamic history as "the triple choice" — conversion, death, or submission — they should be taken seriously. These are all issues that could be identified as markings of further attacks against America.

Ahmadinejad's Bully Pulpit

September 20, 2006

> *"We are in the process of a historical war between the world of arrogance and the Islamic world."*
>
> — MAHMOUD AHMADINEJAD, OCTOBER 26, 2005

According to his official Web site, Mahmoud Ahmadinejad has met with more than 30 heads of state in Iran and more than 60 other high-ranking world leaders in 15 countries, as well as with heads of terrorist organizations, since he became president of Iran a little over a year ago.

Over this period, Mr. Ahmadinejad has used his "bully pulpit" to make contentious statements against Jews, Israel, the West, and in particular the "great Satan" — i.e., America — which he describes as hegemonic and as a superpower made of straw. He warns that America will be "slapped,""branded in shame," and "removed from power."

On July 25, 2005, in one of his first speeches after being elected, the president appeared on Iran's Channel 1 and told the world: "The message of the [Islamic] revolution is global, and is not restricted to a specific place or time. It is a human message, and it will move forward. ... Allah willing, Islam will conquer ... all the mountaintops of the world."

One of Mr. Ahmadinejad's global objectives for the world, with Iran at the helm, is to put America in its place."Today, all the peoples have awoken," he said on IRINN TV on August 2. "The Iranian people are the standard-bearer of this awakening. ... From the southernmost point in South America to the easternmost point in Asia, all the people are shouting a single cry, with placards in their hands and clenched fists."

In a warning broadcast on IRINN TV two weeks later, he added, "You should acknowledge ... the might of the Iranian people, and you should bow and surrender. ... If you do not accept this, the Iranian people will force you to bow and surrender."

The Iranian Students News Agency published a speech by Mr. Ahmadinejad at the World Without Zionism conference last October in which he cited the disappearance of three regimes —the shah's in Iran, Saddam Hussein's in Iraq, and the Soviet Union — as precursors for the destruction of America and Israel. "Is it possible for us to witness a world without America and Zionism? ... This slogan and this goal are attainable, and surely can be achieved," he said.

"If anyone plans to violate the rights of the Iranian people, the Iranian people will brand a shameful mark of remorse on his forehead," Mr. Ahmadinejad warned in a speech on IRINN TV on March 8. Two months later, he gave an ultimatum to America: "Today, they face a historic choice ... to surrender to justice and ... to abandon their arrogance. ... If they don't, they will soon face the powerful slap of the peoples who have arisen." Mr. Ahmadinejad attacked "the superpower made of straw" — i.e., the American government — in an appearance on IRINN TV on February 1. "In the near future, Allah willing, we will put you to trial in courts established by the peoples."

He elaborated in an August 1 speech on IRINN TV. "In light of America and England's behavior ... they themselves are guilty and criminal, and must be placed on trial. ... I warn them: Know that the fire of the wrath of the peoples is about to erupt and overflow. ... They will drag everybody to the defendant's bench, and will remove them from the throne of power," he said.

Mr. Ahmadinejad often uses his "bully pulpit" to explain that Islam is far superior to Western liberalism. In a speech on January 5 on IRINN TV, he said: "One of our slogans during the revolution was 'we will convert the entire world to Islam without logic.' We are confident that the Islamic logic, culture, and discourse can prove their superiority in all fields over all schools of thought and theories."

Iran's Fars News Agency reported on his address to leaders of the Non-Aligned Movement in Cuba on Sunday, "Ahmadinejad described [Western] liberalism as a failed school of thought, experiencing its demise, and added that world hatred for the arrogant attitude and expansionist policies of the U.S. is on the increase, while the U.S., which is the symbol and manifestation of imperialism, is moving on a steep slope toward annihilation."

As the world body gathers in New York this week, a question that must be asked is: Will anyone stand up against this bully Ahmadinejad?

The Verse of the Sword

September 27, 2006

> *"Islam, as portrayed in the encyclopedia of Islam, in the Koran, and the Hadith, was spread by means of the sword. The sword played a major role in spreading Islam in the past, and it is the sword that preserves Islam today. Islam relies upon jihad in spreading the religion. ... Spreading Islam by means of the sword is a duty incumbent upon all Muslims."*
>
> — COPTIC PRIEST ZAKARIYA BOUTRUS, AL-HAYAT TV, JUNE 16, 2005

The Austrian-Iraqi Association for Development held its first conference on extremism and terrorism in January. In attendance were Arab, Muslim, and European researchers, intellectuals, and diplomats, including Kamel Al-Najjar.

Mr. Najjar authored an article for the reformist Web site Elaph.com on February 3 in which he expressed reservations about the validity of recommendations from such conferences. He wrote: "If the Muslims are serious about presenting the radiant face of Islam, they must drop this false, hollow arrogance and acknowledge their dark past and their even darker present. ...They must also reinterpret the Koran, putting the warlike verses — the verse of the sword and similar hadiths — in historical perspective as verses that are no longer valid."

The issue of Koranic verses supporting the spread of Islam by the sword has been making headlines since Pope Benedict XVI's contentious speech on September 12.

As examples from Saudi Arabia, Egypt, Indonesia, and Iran highlight, figures within the Muslim world have debated this subject, with some calling for such verses to be invalidated and others explaining why they are relevant and to be used.

In April 2004, representatives of 65 countries convened in Egypt for the 16th International Islamic Conference, which was dedicated to "Tolerance in the Islamic Culture." Leading religious figures participated in the conference, including the then-Palestinian Authority mufti, Sheikh Ikrima Sabri, who stressed that Islamic history is full of examples of tolerance toward other religions.

In an article in the London Arabic daily Asharq Al-Awsat on May 2, 2004, Mr. Najjar, who was also in attendance, reported that at this and other conferences, leading Islamic figures often quote peaceful Koranic verses, such as "There shall be no compulsion in religion."

He noted, however, that many of the same imams who participated in the Egyptian conference had in the past "insisted that these [peaceful] verses are abrogated by the 'verse of the sword.'"

One such figure is the influential Saudi cleric Musa Al-Qarni, who once was considered Osama bin Laden's mufti. During the Saudi royal family's counterterrorism conference in February 2005, in which he was a participant, he appeared on Saudi IQRA TV to discuss the spread of Islam by the sword.

"We must realize that Allah obligated us to disseminate this religion all over the globe," he said. "And first, it should be spread through outreach and calling people to Allah's word, through pleasing words. ... However, if we run up against someone who opposes this path and attempts to obstruct the spread of the upright religion, and to obstruct their reaching others, in this case it is a duty to fight such a person."

Mr. Qarni, who frequently appears in the Saudi press, added, "We don't agree with those ... who say that the religion [of Islam] doesn't use the sword. No. Islam uses the sword when there is no other alternative. ... If there is need for the sword, then it is wise to use the sword."

"Islam is in two parts, the Koran and the sword," an Indonesian Islamist cleric with a considerable following, Abu Bakar Bashir, said August 21 in an interview posted on Al-Jazeera's English-language Web site. "The Koran has all the guidelines, rules, norms, laws, and punishments we need. The enforcement of the Shariah is the sword we are talking about."

Praise of the sword within Islam is not new, however. In June, an influential Web log, Little Green Footballs, posted excerpts from a 1942 speech by Ayatollah Khomeini in which he stated: "Islam says: Kill the [the non-Muslims], put them to the sword, and scatter [their armies]. ... People cannot be made obedient except with the sword! ... There are hundreds of other [Koranic] ... and hadiths ... urging Muslims to value war and to fight. Does all that mean that Islam is a religion that prevents men from waging war? I spit upon those foolish souls who make such a claim."

On Monday, Pope Benedict met with Muslim diplomats at the Vatican to try to ease the tensions that ensued after his mid-September speech.

As the meeting took place, calls continued for the pope to convert to Islam and for crosses to be broken, and Jesus was taunted as "the slave of Allah."

Those denouncing such statements in the Muslim world have been few and far between.

Pope Benedict's Islamic Conspiracy

October 4, 2006

Controversies in the Middle East are surrounded by conspiracy theories that proliferate in the region's press. Recent conspiracies have involved the identity of the person behind the cartoons of Islam's Prophet Muhammad, the genocide in Darfur, and the true perpetrators of terrorist attacks worldwide.

The latest Middle East conspiracy was hatched September 12, following Pope Benedict XVI's contentious speech in Germany on Islam. Influential Middle East press and broadcast outlets and prominent figures claimed that America, Jews, Zionists, and/or Israel were really behind his comments.

The Iranian judiciary chief, Ayatollah Mahmoud Hashemi Shahroudi, said in a meeting with judiciary officials September 18 that the Zionists — backed by America and Europe, "which consider themselves Christians" — were behind the pope's statements, according to the Islamic Republic News Agency.

The Iranian daily Jomhuri-ye Eslami reported on September 17 that the pope's remarks against Islam could have been "a dictate that the Zionists and the Americans have written [for him] and have submitted to him."

The Kayhan daily, which is closely allied with the Iranian supreme leader, Ayatollah Ali Khamenei, reported on September 17, "There are many signs that show that Pope Benedict XVI's remarks regarding the great prophet of Islam are a link in a connected chain of a Zionist-American project."

Ayatollah Khamenei was quoted by the Mehr News Agency as saying the pope's remarks were part of an American-Israeli conspiracy to create a conflict between religions.

He added that America is trying to sow discord between Muslims and Christians, tarnish the image of Muslim minorities in the West, and suppress Islamic countries under the pretext of their alleged support for terrorism.

In a September 17 Tehran Times op-ed titled "The Pope's New Crusade,"Hassan Hanizadeh wrote: "Following the suspicious 9/11 attacks, the West is increasing the psychological and propagandistic pressure on Muslims to prepare the ground for the expulsion of Muslims from the U.S. and European countries.

"The Pope's rude remarks are actually code words for the start of a new crusade. ... Another aim of such propaganda is to facilitate the implementation of the new Middle East plan so that the Zionist regime can become the Hong Kong of the Middle East and fill Islamic countries' markets with Israeli goods."

High-ranking Saudi officials fear that the pope's remarks might represent the beginning of a new papal role in directing efforts against Islam, the Arab reformist Web site Elaph, which is bankrolled by Saudi money, reported September 15.

The article said the pope is a believer in white supremacy and Christianity as the only civilized religion, and it also suggested that his statements could be an attempt to forge closer ties with Israel.

A renowned Islamic scholar, Mufti Abul Irfan Mohammad Naimul Haleem, issued a statement on September 16 that was reprinted in the Indian newspaper the Hindu.

In it, the mufti said Western countries had "hatched conspiracies and launched a disinformation campaign against Islam and Muslims to divert the world's attention from the expansionist designs of some of them against certain sovereign Muslim countries."

The pope's remarks, he added, were "part of meticulously planned propaganda against Islam and the Prophet Mohammad to create hate and mistrust against Muslims."

The Turkish press reacted to Benedict's remarks with attacks on Jews and Christians; one influential writer called them "blood sucking vampires."

A columnist for the secular, centrist daily Aksam, Hüsnü Mahalli, wrote on September 19: "Contrary to some people's interpretations, the pope has spoken knowingly, consciously, and deliberately, having calculated all the consequences. ... On his first visit to his homeland Germany [as pope], 'his majesty' visited a Jewish synagogue, and there, in a secret meeting with 60 rabbis, discussed how they should act together against Islam."

When the next major event occurs in the Muslim world — whether it is manmade, natural disaster, or something else — the Jews, Zionists, and America can expect to be made the scapegoat for it.

The Influence of a Man Who Denies the Holocaust

October 12, 2006

> *"There was never a single execution gas chamber under the Germans.... So all those millions of tourists who visit Auschwitz are seeing a lie, a falsification."*
>
> — ROBERT FAURISSON, SAHAR TV, FEBRUARY 2005

He exposed the "big lie of the alleged Holocaust." He proved that "the 'Diary of Anne Frank' [is] a fraud" and that "the gas chambers were fabricated."

When Arab and Iranian TV networks such as Al-Jazeera and Sahar need someone to discuss the "big lie of the alleged Holocaust," they go to him.

When Arab and Iranian politicians such as the Palestinian Authority leader, Mahmoud Abbas, and President Ahmadinejad of Iran want to provide evidence that no "alleged crimes" were perpetrated against Jews in World War II, they cite him.

Like other European Holocaust deniers who enjoy rock star status in Iran and the Arab world, Robert Faurisson first appeared on the scene in the 1970s, when he refuted accounts of the gas chambers and the Nazis' systematic killing of Jews.

"The dean of deniers," as Mr. Faurisson is known, is a sought-after interview subject in the Middle East. When Al-Jazeera devoted a program to a planned international conference of Holocaust deniers in Beirut in May 2001, it called Mr. Faurisson.

"We have proved and are still proving that there was no massacre or Holocaust of the Jews, and that there were no gas chambers for the Jews, and that the figure of 6 million victims is exaggerated ... saying the truth about the biggest lie of the 20th and 21st centuries, the lie of the Holocaust," he said.

Arab and Iranian "experts" on the Holocaust frequently cite Mr. Faurisson's theories. In his 1982 doctoral dissertation at the Peoples' Friendship University in Moscow, Mr. Abbas discussed "the secret ties between the Nazis and the Zionist movement leadership." Two years later, the Jordanian publisher Dar Ibn Rushd put out an Arabic-language book based on Mr. Abbas's dissertation.

"Regarding the gas chambers, which were supposedly designed for murdering living Jews: A scientific study published by Professor Robert Faurisson of France denies that the gas chambers were for murdering people, and claims that they were only for incinerating bodies, out of concern for the spread of disease and infection in the region," Mr. Abbas wrote.

In January, when Mr. Ahmadinejad described the Holocaust as "a myth," Mr. Faurisson sent him a letter "expressing his full support of his remarks." The letter was heralded in the Iranian press.

On September 20, Iran's IRINN TV broadcast a report on the country's ongoing Holocaust cartoon contest. The curator of the museum exhibiting the cartoons, Masoud Shojai Tabatabai, told the station that the display proves that the "alleged crimes" that occurred at places like Auschwitz "are in fact a lie." The curator said the proof was based on "the very serious and accurate analysis of Mr. Robert Faurisson."

The editor of the conservative Iranian daily Kayhan, Hossein Shariatmadari, wrote an article on December 13, 2005, in which he cited Mr. Faurisson's lectures at the University of Lyon and one of his books, "The Gas Chambers: Reality or Legend?"

The book, translated into Persian by Seyyed Abu Al-Farid Zia Al-Dini, examines dozens of documents "where the Zionists claim" the slaughter took place, "such as the gas chambers, the fabricated museum of the crematoria, Dachau in Munich," Mr. Shariatmadari wrote. Mr. Faurisson "conducted precise and scientific conversations with hundreds of witnesses and ultimately showed, with no interpretation and by means of documents only, that the affair of the slaughter of the Jews in Nazi Germany is a great historic lie."

When the United Nations announced that it would designate a day to commemorate the Holocaust in November 2005, Mr. Faurisson gave an interview to the Tehran Times.

"For many years now, I have been telling my acquaintances in the Muslim world that the Jews and the Zionists want to impose the religion of the alleged 'Holocaust' of the Jews on the whole world," he said. "... The Muslim world has been awakening from its too long torpor for only a few years. It ought to have listened to the revisionists long ago and denounced out loud the sham of an alleged German project to exterminate the Jews, the alleged Nazi gas chambers, and the alleged 6 million Jewish victims."

On October 3, a French court convicted Mr. Faurisson of Holocaust denial for statements he made on Iran's Sahar TV in February 2005. It was the fifth time he has been found guilty for "complicity in contesting the existence of a crime against humanity."

An Islamic TV Channel Expands Its U.S. Audience

October 19, 2006

Bridges TV, an American-Islamic TV channel "seeking to improve the image of Muslims in the United States" and to "offer a unique perspective on the Middle East and the war on terrorism," has extended its availability into six states, creating a potential audience of nearly 2 million.

The network's programming includes a mix of entertainment, sports, news, documentaries, and advertisements from companies like Ford, with an emphasis on religious programs.

The channel says it has been endorsed by "top American [Islamic] scholars and community leaders," whose representatives appear on many of its programs, including one called "Prominent Scholars."

Some speakers openly criticize Islamic extremists. An imam from Los Angeles, Sheikh Tajuddin Bin Shuaib, appeared on the channel on October 8 and denounced Osama bin Laden and the September 11, 2001, hijackers.

Some guests, however, are extremists. One religious figure who appeared October 3 said Muslims have a duty to change America and to increase their numbers to 50% of the population from 2%. He recommended that Shariah, or Islamic law, be implemented in American courts.

During a roundtable discussion on the Arab-Israeli conflict on October 5, one participant offered a solution: "For the Jews to leave and return to Europe."

Bridges TV aired a speech by the influential Muslim scholar Jamal Badawi on October 4. Mr. Badawi, who teaches Islam throughout North America, gave an interview to the Saudi Gazette on June 24, 2005, in which he raised questions about who was behind the September 11 attacks and suggested that Americans could be behind the car bombings of Iraqi markets.

Every night, Bridges TV shows a news program, "Talking Points." Its guest on October 4 was Imam Mohammad Alo Elahi, whom it described as a leading "interfaith figure." According to his Web site, Imam Elahi was a spiritual leader in Ayatollah Khomeini's Iranian navy and also is the leader of "one of the largest mosques in the U.S.," in Dearborn, Mich.

The Web site describes his meetings with world leaders and shows photographs of him with the spiritual adviser of Hezbollah, Ayatollah Muhammad Hussein Fadlallah; Ayatollah Khomeini; Presidents Khatemi and Rafsanjani of Iran; Secretary-General Annan of the United Nations; and Minister Louis Farrakhan.

Throughout the day, Bridges TV airs segments of Koranic verses, quite a few of which denounce "unbelievers." One notable verse that aired October 9 praised martyrdom.

Since the Islamic holy month of Ramadan began, the channel has been showing official, Saudi government-controlled Wahhabi sermons from Mecca's holiest mosque, Al-Haram. The sermons stream live via Saudi TV Channel one every day at 4 p.m., and Bridges TV adds its own English subtitles.

An anti-Jewish, anti-Christian sermon from October 5 included the call, "May God destroy them!"

Bridges TV's Web site, bridgestv.com, features a weekly poll. Notable questions and results include 59% calling for Hezbollah to continue as a military force in Lebanon, 73% in agreement with the American policy of withholding funds to the Hamas-led Palestinian Arab government, and 63% believing that the Iranian government's nuclear program is for peaceful purposes.

One of the stars of Bridges TV is a cofounder and vice chairman of the international health care company CBay Inc., Donald "Skip" Conover, who hosts and produces a show called "Words Matter." He was the subject of a gushing article in the Saudi daily Arab News on September 27.

In the article, Mr. Conover expressed "his disgust" at what he called inflammatory statements about Arabs and Muslims in the press.

He also discussed the power of the "Jewish lobby" and called on all Muslims to vote for the Democratic Party. "Every American politician is in lockstep with Israel. … If they vote against, then the Jewish lobby will put a lot of money behind the candidate against them in their districts in the future. I have news for the Muslim community. All American politicians are in the pocket of the Jewish lobby today because they control a lot of money, and they spend a lot of money in politics."

"If the Muslims of America believe that they don't want Bush to have a free hand for the next two years, then the Muslims of America need to get organized and make sure they get out to vote for Democrats for both the House and the Senate," Mr. Conover added. "Every Muslim in the Middle East who has a relative in the U.S. should get the message across to their relatives. They need to make sure that all their friends vote against Bush."

Bridges TV claims that its "major purpose" is "to build bridges between American Muslims and other Americans." After viewing the channel, I find this highly unlikely.

Western Press Ignores Iran's Hate-Filled Quds Day

October 25, 2006

It is disturbing when the entire leadership of one nation, along with hundreds of thousands of its citizens, comes out with celebrations and parades every year that call for the annihilation of another country.

It is more twisted that no world leaders or international bodies, including the United Nations, have denounced the activities surrounding Quds Day, an Iranian holiday introduced by Ayatollah Khomeini that is marked on the last Friday of Ramadan.

While the world was focusing on North Korea last week, Iran's mad scientists were hard at work preparing for the annual Quds Day celebrations.

Most of the Western press outlets that reported on the popular holiday simply downplayed it as just another "anti-Israel" day. However, this year's revelries focused both on calling for the annihilation of America and embracing Iran's nuclear program.

The celebrations included proclamations by the country's leaders and activities for university students and artists.

Isfahan University's Mechanical Energy College took first place in a Quds Day competition for its design of a pilotless plane that can be used for "suicide attacks." The director of the Iranian Broadcasting Organization of Music Production, Mohammad Mirzamani, composed a symphony dedicated to "the victory over the Zionist regime," and the country's religious Web logs were told to report on all the festivities.

Iranian press outlets featured hundreds of photographs from the celebrations in Tehran. Among the notable scenes captured were children in Condoleezza Rice costumes; effigies of President Bush, Prime Minister Olmert, and Prime Minister Blair being lit on fire and dragged through the streets; the burning of American and Israeli flags; and hundreds of posters of Sheikh Hassan Nasrallah featuring the caption "I swear to Allah that Israel is weaker than [a] spider house." The posters called for a boycott of such "Israeli" goods as McDonald's, Kit Kat bars, Intel, L'Oreal, Nestlé, Disney, and Marlboro.

President Ahmadinejad gave a series of speeches leading up to and on Quds Day. At an Iftar address on October 14, he discussed his "connection with God" and said: "The president of America is like us. That is, he too is inspired ... but [his] inspiration is of the satanic kind. Satan gives inspiration to the president of America."

Mr. Ahmadinejad delivered his Quds Day speech under a banner that read, "Israel must be wiped off the face of the world." He described the holiday as "a day for confrontation between the Islamic faith with the global arrogance."

In another speech, he said Israel was "doomed" and promised that the Israeli "regime will be gone, definitely."

The words "the Zionist regime is a cancerous gland that needs to be uprooted" were written in a communiqué from the Iranian Foreign Ministry in honor of the holiday. Foreign Minister Manouchehr Mottaki held a meeting for other Islamic countries' ambassadors to Iran and told them that Israel's existence would be shattered and that death bells were tolling for the Zionists. At the meeting, the Palestinian Arab ambassador to Tehran, Salah Zawawi, said, "The day for the liberation of Quds Day is close at hand."

A who's who of the Iranian leadership marched in the main Quds Day parade before crowds chanting "death to Israel" and "death to America." The marchers included a former Iranian president, Mohammed Khatemi, and a spokesman for the parliament presidency board, Mohsen Kouhkan, who predicted a quick "final and total defeat of America and the Zionist regime."

The chief of the judiciary, Ayatollah Mahmoud Hashemi Shahroudi, praised "the fasting people taking part in the rally [who] are chanting slogans such as 'death to America' and 'death to Israel.'"

"The world arrogance and Zionism today are shivering from Muslim vigilance and are on the threshold of annihilation," he added.

Information Minister Hholam-Hossein Mohseni Ejei explained that the holiday "is a proper occasion for people to declare their hatred of America and Israel," while a representative of the Islamic Consulate Assembly, Ahmad Pish-bin, promised that the "final defeat for world arrogance" is coming.

The chairman of the Expediency Council and a former Iranian president, Akbar Hashemi Rafsanjani, who led Friday prayers, said Quds Day is an important factor "between Islam and unbelief and the stage for Muslims' jihad." He added, "The world's 1.5 billion Muslims back this jihad."

Mr. Rafsanjani also led Quds Day prayers on December 14, 2001. Then, he warned of a coming confrontation between the "pious and martyrdom-seeking forces" and the "highest forces of colonialism," which "might inflame a third World War."

Sadly, Mr. Rafsanjani is considered one of Iran's more moderate leaders.

Arabs and Iranians Root For the Democrats Again

November 1, 2006

> *"The Muslims of America need to get organized and make sure they get out to vote for Democrats for both the House and the Senate. ... Every Muslim in the Middle East who has a relative in the U.S. should get the message across to their relatives."*
>
> — SKIP CONOVER, HOST ON MUSLIM-AMERICAN BRIDGES TV, ARAB NEWS, OCTOBER 19

As the November 7 midterm elections approach, Iranian and Arab officials and journalists are hoping the Democrats will win control of the House and Senate.

It should be noted that before the 2004 elections, they also overwhelmingly predicted and rooted for a Democratic victory:

- "I would like to appeal to you, Muslim citizens and voters of America ... cast your vote against Bush ... in the name of Islam," a former prime minister of Malaysia, Mahathir Mohamad, said.
- "We advise the American people ... do not vote for Bush and his ilk," Ayatollah Javadi-Amoli of Iran said.
- "Arab-Americans should rethink their alliance with Bush," an Arab-Israeli member of the Knesset, Azmi Bishara, said.
- "The most important thing in the coming elections is if Kerry wins and neo-conservatives leave," the Lebanese minister of information, Michael Samaha, said.
- "Bush, Allah willing, will go to hell in this coming November," the editor of the Egyptian newspaper Al-Gil, Nagi Al-Shihabi, said.

Fast-forward two years, to the current election cycle. Similar calls are being heard, and some articles have even suggested that President Bush may "wag the dog" with an attack on Iran to help the Republicans.

"The Republican Party is in so much trouble it may risk a major confrontation in the Middle East before half-term congressional elections in November. Republicans are far behind in the polls," an editorial in the October 19–25 Egyptian state-run English-language weekly Al-Ahram said.

"If the Democratic Party seizes Congress or the Senate. ... This might even open the door for impeachment," the columnist Sandy Shanks wrote on Al-Jazeera's Web site on October 22. He asked rhetorically, "How badly does Bush want to win next month?"

"Will Bush attack Iran by air and sea before next month's elections to prove to his domestic audience that his party is best at preserving and protecting our way of life, thereby 'ensuring' a Republican victory in Congress?" Mr. Shanks added.

The August 30 edition of the London-based Arabic-language daily Asharq Al-Awsat featured "The U.S. Republican Party and Defeat," an article by an Al-Arabiya TV host and columnist, Hussein Shobokshi. "The Republican Party is beginning to see the danger signs and recognize that the current administration will cause it to lose ... and suffer a defeat in the next elections," he wrote.

On Sunday, a group of Arab diplomats that included the leaders of Hamas and Islamic Jihad and Foreign Minister Manouchehr Mottaki of Iran met in Damascus. Among the topics of discussion was the American elections. Mr. Mottaki called Mr. Bush a "loser" and said that after next week's elections, the president will be politically paralyzed and unable to initiate any political moves.

As he discussed the upcoming elections before a crowd chanting "death to America, death to Israel," Hojatoleslam Ahmad Khatami, who led last week's Friday prayers in Tehran, said: "The savage and devilish acts and performances of Bush, as a Republican, have placed his party in a very critical condition."

An editorial in the October 27 Saudi English-language daily Arab News said, "As the hours tick by to the November 7 midterm election, the Bush administration is wiggling frantically on the butcher's hook." The editorial criticized American actions in Iraq and Mr. Bush for "trying to bolster electoral support for his Republican Party at the ballot box." It added, "Bush went gunning for the Iraqi bad guys who had fooled his pa. He did so in the name of his vengeful war on international terror. He was eagerly supported by an angry America."

The Kuwaiti daily Al-Seyassah's October 29 "Digest of Public Opinion" discussed the midterm elections, saying that if "extremists such as Bush, Cheney, Rumsfeld, Rice, and the Republicans at large rule longer, America will likely be transformed into a dictatorship and as such we expect that the situation currently existing in Guantanamo is likely to face the Americans themselves. ... Consequently, most political observers are currently expecting the Republicans to incur an overwhelming defeat during the upcoming congressional elections."

One must ask why so many press outlets and Arab and Iranian leaders that are known to be vehemently anti-American are openly supporting one political party over the other.

A New Al-Jazeera and 9/11

November 8, 2006

> *"... Men like bin Laden will not allow the Islamic world to bow down under the infidel enemies' tyranny, under the tyranny of America. ... Osama bin Laden is a jihad warrior who implemented the principles of Islam. ... Allah guarded his soul until he reached a point where he became a hero."*
>
> — SAUDI MINISTRY OF ISLAMIC AFFAIRS PREACHER ABDALLAH BIN MATRUK AL-HADDAL, AL-JAZEERA, JANUARY 2002

Just in time for its 10th anniversary, Al-Jazeera is reportedly set to launch its English-language network November 15. On that day at exactly 3 p.m., Prime Minister Blair is scheduled to take a live tour of Al-Jazeera International's London bureau and sit down for an interview with Sir David Frost.

The 100-person Al-Jazeera International bureau includes an impressive staff of former CNN, CNBC, BBC, AP TV, and ABC employees. "The launching of the English channel offers the chance to reach out to a new audience that is used to hearing the name of Al-Jazeera without being able to watch it or to understand its language," the director-general of Al-Jazeera, Wadah Khanfa, said recently.

Al-Jazeera International's head of news, Steve Clark, told a New Zealand newspaper on May 3 that "Al-Jazeera International will share resources and facilities with its Arabic sister channel, and the two broadcasters will work together on relevant stories."

When Mr. Clark was asked how a Palestinian Arab suicide bombing in a crowded Israeli marketplace would be reported, he answered, "We don't condone acts of terror, but we'd be very careful about labeling particular groups as terrorists."

Over the next few weeks, this column will focus on the controversy surrounding the AJI launch. To understand the true intentions of the network, it is important to examine what Al-Jazeera was broadcasting leading up to and shortly after the attacks of September 11, 2001.

One of the most popular shows on Al-Jazeera, "Opposite Direction," is hosted by Dr. Faysal Al-Qassem, who could be described as the network's Larry King. Dr. Qassem is featured prominently on Al-Jazeera's 10th anniversary Web page.

On July 10, 2001, Dr. Qassem devoted an entire program to "Bin Laden — Arab Despair and American Fear." As he introduced the show, he said: "Bin Laden has made the greatest power in history shudder at the sound of his name. ... Has bin Laden not become a worthy opponent, feared by America? ... Who ... caused its troops to run like rabbits? ... Who has recently become the no. 1 Arab and Islamic hero? Does the U.S. fear him because it sees him as a terrorist, or because he is the conscience of the Arab and Islamic world?"

Another program, from May 23, 2001, was devoted to suicide bombings and how they could be used against America. Many of the notable Arab figures who called in to the May 23 show made threats against American interests. The mufti of the Palestinian Liberation Army, Nader Tamimi, called in and said: "Your hands are blessed when you kill the enemy, blessed be you in heaven, with the Prophet Muhammad. ... What have the martyrdom operations achieved so far? They have removed NATO from Lebanon, the American forces from East Saudi Arabia ... the American forces from Somalia. My blessings to those who carried out the [USS] Cole operation."

One might have thought that after September 11, Al-Jazeera would have toned downed its incitement to violence against America, but as this column will show in the coming weeks, it did not.

In Fouad Ajami's must-read article on Al-Jazeera in the New York Times magazine of November 18, 2001, he eloquently explained: "Al-Jazeera ... may not officially be the Osama bin Laden Channel — but he is clearly the star. ... The channel's graphics assign him the lead role: there is bin Laden seated on a mat, his submachine gun on his lap; there is bin Laden on horseback in Afghanistan, the brave knight of the Arab world. A huge, glamorous poster of bin Laden's silhouette hangs in the background of the main studio set at Al-Jazeera's headquarters."

Mr. Ajami also issued a warning about Al-Jazeera that has gone unheard and should be reconsidered as AJI officially opens the doors of its fancy K Street office: "Al-Jazeera's virulent anti-American bias undercuts all of its virtues. It is ... a dangerous force. And it should treated as such by Washington."

Al-Jazeera in English, Part II

November 15, 2006

"I say to Iraqis, whoever cannot slaughter [a sheep] ... should take an American soldier and slaughter him."

— ABD AL-KARIM ABD AL-RAZZAQ,
AL-JAZEERA, NOVEMBER 22, 2005

On November 25, 2005, Al-Jazeera covered an anti-American demonstration of enraged people cursing America and President Bush.

"Down with fascist America!" "Bush is a criminal! ... He should collapse and be trampled on!" "Al-Jazeera ... exposed Bush and his despicability!" the crowd chanted. Such scenes are not uncommon in the Middle East — but in this case, the demonstrators were Al-Jazeera staff in Ramallah.

Other employees of the network — which, at the last minute, has changed its name to Al-Jazeera English — hold the distinction of being jailed at Guantanamo Bay and around the world for their connections with terror organizations. In an Al-Qaeda Internet news broadcast last year, an anchor even expressed solidarity with his jailed "brother," who works for the network.

In a June 25 Op-Ed for the Washington Post, "Al-Jazeera, as American as Apple Pie," the executive producer of programming for Al-Jazeera English, Joanne Levine, said many Americans unfairly view Al-Jazeera as "terrorist TV" or "Osama bin Laden's network." Those who work for AJE face "prejudice" from Americans, she said, and she praised the network for having the "courage to tell it how it is."

Members of AJE staff have been making the rounds in the press to spin the positive aspects of the channel, even hiring a former Clinton administration figure who served as a consultant to the State Department, Michael Holtzman, to handle its public relations efforts.

A former "Nightline" correspondent, David Marash, discussed the "intellectual integrity" of the new channel in an April 3 column for the New York Daily News, "Why I Joined Al-Jazeera." And the managing director of AJE, Nigel Parsons, called critics of the channel "hypocrites." "How you see something ... depends very much on where you're sitting," he said in the March 26 edition of the New York Times.

The esteemed BBC host David Frost, who is currently working for AJE, has stated in multiple interviews that the West's perception of the channel is incorrect. "Obviously, we all suffer from the handicap of not being able to sit there and watch in Arabic," he told the Washington Post in October 2005.

Some of the most vocal anti-American world leaders have already embraced AJE. When the launch of the station was first announced, a British member of Parliament, George Galloway, appeared on Al-Jazeera to give his endorsement.

President Chavez of Venezuela visited the station's studios in Qatar on August 4. "Whenever I meet an Al-Jazeera reporter, I feel like giving him an interview," Mr. Chavez said. "Allow me to say to the Al-Jazeera staff that I am amazed by them," he added. "It is an honor for me to be here ... in these studios, and among these ... brothers in Al-Jazeera. ... You are an example and a model of honor, bravery, and the fight for justice. You are committed to the code of journalistic work ... and the distortion of the truth, despite all the pressure."

One Al-Jazeera broadcast that Messrs. Galloway and Chavez would likely have enjoyed was a February mock trial of President Bush, Prime Minister Blair, and Prime Minister Sharon. The trial, which the channel covered extensively, was sponsored by a group of influential Arab leaders. "This is not a mock trial. This is a trial in absentia," the head of the tribunal and a former Malaysian president, Mahathir Muhammad, told Al-Jazeera.

The deputy secretary-general of the Union of Arab Lawyers and a member of the Egyptian Parliament, Abd Al-Azim Al-Maghrabi, said the trial was created to challenge "America's right to subjugate humanity ... as an absolute dictator that has the right to annihilate whoever strives for freedom."

In an October 31 article on Al-Jazeera's Web site, the director-general of AJE, Wadah Khanfar, described the channel as being the "voice of the voiceless." But do such voices include notable Arab figures who call for attacks against America, such as the head of the Egyptian Labor Party, Magdi Ahmad Hussein, who called for the bombing of Los Angeles? What about Al-Jazeera's daily broadcasts of footage of American soldiers in Iraq being killed, which air as the channel's hosts visibly strive to hide the smirks on their faces?

A platform for such hatred should not be welcomed, and certainly not from an office located a few blocks from the White House.

Americans Should Beware Of the Sheikh of Al-Jazeera

November 22, 2006

> *"O God, destroy the Zionist, the American, and the British aggressors. O God, shake the ground under them."*
>
> — SHEIKH YOUSUF AL-QARADHAWI, UMAR BIN-AL-KHATTAB MOSQUE, DOHA, QATAR, MARCH 7, 2003

As the new Al-Jazeera English network strives to separate itself from its mother channel, its American employees should denounce their co-worker Sheikh Yousuf Al-Qaradhawi and his statements against their country.

Sheikh Qaradhawi is considered Al-Jazeera's religious authority. He has hosted one of the network's most popular shows since its early days, "Al-Shar'ia wa Al-Haya," or "Islamic Law and Life." The program airs every Sunday at 9:05 a.m. Mecca time, and for 90 minutes focuses on modern life from an Islamic perspective.

Bootleg videos of old episodes of Sheikh Qaradhawi's show are sold throughout the Muslim world. As Hugh Miles wrote in his seminal book "Al-Jazeera," "Throughout the internet and his Al-Jazeera TV slot, the Sheikh influences hundreds of millions of Muslims all over the world."

Sheikh Qaradhawi wears many other hats. He established the department of Islamic research at the University of Qatar and is the founder and head of the European Council on Fatwa and Research, as well as serving as a top spiritual leader for many Sunni Muslims.

The sheikh is also influential in Europe and America, which he has been barred from entering since 1999. While his European Council on Fatwa and Research has its headquarters in Ireland, last year E.U. Parliament members wrote a letter calling for him to be banned from Europe. In the letter, two British lawmakers, Dr. Charles Tannock and Nirj Deva, wrote: "We cannot accept him into our midst and provide legitimacy for those who incite terrorist violence in some places simply because they denounce other types of violence targeting Western cities."

In response, Sheikh Qaradhawi and his supporters have tried to create an image of him as a moderate leader and a victim of a Zionist-American conspiracy. "In the last few months, a war was launched against yours truly. It is a ruthless media war, spanning most of the world in the East and the West. ... This war was waged by global Zionism and its strategic partner, America," Sheikh Qaradhawi said in a Friday sermon on January 7, 2005. "They recruited the soldiers and used mercenary writers and the reprehensible TV channels to attack me. They unleashed rabid dogs here and there to bite me with their fangs and bark at me with their voices. ... I have become the no. 1 inciter for terrorism in the world. ... I have become a preacher of violence."

In an interview with the Arabic-language newspaper Al-Hayat on February 19, 2005, Sheikh Qaradhawi even said America or the Mossad could assassinate him, and that he wanted to die as a martyr.

The sheikh became the focus of criticism within the Muslim world early this year after the controversy surrounding the Danish publication of cartoons of Islam's Prophet Muhammad, when he told his followers through Al-Jazeera: "The nation must rage in anger. ... We are not jackasses for riding, but lions that roar."

During the last week of September, Sheikh Qaradhawi gave a Friday sermon in response to the pope's statements on Islam, which aired on Al-Jazeera. The sheikh is shown standing besides a poster in Arabic that reads,"Down with Christian Zionism. "The poster also shows President Bush walking an animal that looks like a combination of a pig and a dog, which is shown barking the word "baba," which means "pope" in Arabic.

Sheikh Qaradhawi's venomous hatred of the West and anti-Americanism takes many forms, including calls to attack America, boycott American products, and even kill Americans in Iraq.

Just another reason why Al-Jazeera has a bad reputation.

Al-Jazeera's Dirty Little Secret [Unreleased]

November 29, 2006

"Al-Jazeera Satelitte Network, the Greatest Arab Media Organization."

— IKHWAN WEB.COM
[THE WEBSITE OF THE MUSLIM BROTHERHOOD], OCTOBER 25, 2006

Within the Arab and Muslim world, a dirty little secret is widely known and openly discussed about the Al-Jazeera network. It has yet to be reported on the Western media, but it is of utmost importance in understanding the true motives behind the channel.

Pierre Akel, founder and editor of the influential Arabic reformist website Metransparent.com best summed it up on February 9, "When it comes to satellite television in the region [Middle East], Al-Jazeera is controlled by the Muslim Brotherhood."

Arab reformists who know first hand about the danger of the network often discuss this connection.

Tunisian intellectual, Dr. Khaled Shawkat, director of the Netherlands–based Center For Promoting Democracy in the Arab World wrote about it in an article for the progressive Arabic website Elaph.com on July 29.

In addition to criticizing the channel for "stirring up emotion on sensitive regional issues," he explained: "Al-Jazeera has been hijacked by the Muslim Brotherhood organization – either at the wish of the channel's owners as part of a certain political game [played] by the Qatari rulers, or out of the lack of awareness..."

In an article in Metansparent on December 22, 2004, Arab reformist Abu Khawla also discussed the Muslim Brotherhood's connection to Al-Jazeera, including the channel's religious authority, Sheikh Yousuf Al-Qaradhawi who as a member of the organization left Egypt forty-five years ago to seek asylum in Qatar.

Khawala wrote, "To obtain credible information, Arabs turn nowadays to satellite TV channels, especially Al-Jazeera... a tool of communications that is so far completely monopolized by fundamentalists. In the case of Al-Jazeera, preacher Al-Qaradhawi (a leading figure of Egypt's Muslim Brotherhood) is one of its most influential founding members. And fundamentalists are in charge of its news programs and talk shows..."

"The satellite channel of the masked terrorists" which "makes great efforts day and night to encourage terrorism," is how Dr. Abd Al-Khaliq Hussein, a contributor to Elaph.com explained how Al-Jazeera was described in some Arab circles on September 5, 2004. The article was highly critical of the channel and focused on its connection to the Muslim Brotherhood, as well as Sheikh Al-Qaradhawi's position there.

The general manager of Al-Jazeera, Waddah Khanfar gave a wide ranging interview to the website of the Muslim Brotherhood about his channel's 10th anniversary and the launch of its English network on October 25. The website seemed to have caught Khanfar off guard when it asked him, "Mr. Khanfar, you have an Islamic background, specifically, the Muslim Brotherhood, don't you think that this is conflicting with your position as a head of the biggest Arab media organization?"

It should be noted that many leaders of the Muslim Brotherhood utilize Al-Jazeera as their own platform for enraging the Arab and Muslim masses against the U.S. and for Jihad. For example, Mahdi 'Akef, Supreme Guide of the Muslim Brotherhood, appeared on Al-Jazeera on July 30, 2006 and stated: "This American Satan claims to be a messenger of divine guidance... I salute all the honorable Americans who stand up to this Satan, who wants to set the entire world on fire, not just the Arab and Islamic nation. I go back to the issue of Jihad. Jihad is an individual duty incumbent upon every Muslim..."

From the General Guidance Council of the Muslim Brotherhood, Dr. Abd Al-Muni'm Abu Al-Futuh, appeared on Al-Jazeera on April 25, 2004, and spoke in favor of armed attacks against American soldiers in Iraq. He specifically stated all members of the Muslim Brotherhood throughout the world want to battle U.S. troops: "Regarding the resistance, I say clearly and without evasion that we support armed resistance. We, as Muslim Brotherhood and as the world leadership of the Muslim Brotherhood, support the armed resistance in Iraq against the American occupier, one hundred percent. I personally, as all members of the Muslim Brotherhood in Egypt and around the world, wish to carry weapons against the American occupation in Iraq..."

Dr. Al-Futuh concluded by stating, "This is a matter of religion, not politics. When a Muslim land is occupied, Jihad becomes an individual duty for every man and woman, boy and girl..."

As Al-Jazeera celebrated its 10th anniversary and launched the Al-Jazeera English network, its successful PR efforts included articles about it in dozens of leading Western newspapers, magazines, and TV segments. Surprisingly, none to my knowledge explored the network's connection to the Muslim Brotherhood.

Mideast Chess Match Plays Out in Lebanon

December 6, 2006

> *"Lebanon, a country that was supposed to become the main base of Western culture in the Arab world, has turned into a base of jihadi and resistance cultures."*
>
> — IRANIAN SUPREME LEADER AYATOLLAH ALI KHAMENEI, JULY 17

Since Friday, the streets of Beirut have been filled with Hezbollah supporters calling for the ouster of the pro-Western government of Prime Minister Siniora. The events currently playing out in Lebanon could be considered the fruits of seeds that were first planted 22 years ago, when the terror organization Hezbollah was founded.

The August 2 edition of this column, "An Islamic Republic Is Hezbollah's Aim," stated that Hezbollah's ultimate goal is "to take over the Lebanese government."

A member of the Lebanese parliament, Mustafa Aloush, was quoted in the Lebanese daily Al-Mustaqbal on September 17 as supporting such a theory. "Hezbollah is a militia with political goals, and first among these goals is the establishment of an Islamic republic in Lebanon, according to the path of the Imam Khomeini," he said.

Many other pieces on the "Middle East Chess Board" also are playing out in Lebanon. Among them, of course, is the America-Iran factor.

On a visit to Iran in the second week of November, the speaker of the Lebanese parliament, Nabih Berri, met with President Ahmadinejad and Ayatollah Khamenei. During his meeting on November 15 with Ayatollah Khamenei, which was covered extensively in the Arab and Iranian press, Mr. Berri was quoted as saying America and Israel would meet with defeat in Lebanon, and that "political developments in the region and the world confirm that a new phase has begun."

"This opportunity must be exploited to the full through determined action and trust in Allah the Almighty," Mr. Berri said. "The latest developments are the beginning of the fall of the American empire. ... The victory of the Lebanese and of the Islamic resistance played a significant part in the defeat of the Republican Party in the recent congressional elections" in America.

Mr. Berri added, "America and the Zionist entity are trying to keep the Islamic Republic from exerting its spiritual influence in the region," but "the Islamic Republic of Iran will be the leading force in the new phase."

Another development playing out now in Lebanon is the Sunni-Shiite dispute. A New York Times report on December 2 detailed how Lebanese Sunnis are supporting the Lebanese government against the majority of Shiites in the country that support Hezbollah.

On this front, Al-Mustaqbal published an article on November 20 by a Sunni sheikh, Said Harmush, that criticized Hezbollah. Members of Hezbollah claim that they are acting in the name of Islam, he wrote, "but they are doing the exact opposite of Islam! ... Roving bands of [Hezbollah members] took over property belonging to 'the Sunni brothers,' as they call them — houses, land, and shops. Then they started with acts of robbery, slaughter, humiliation, and expulsion, much more than the Jews have done anywhere."

Sheikh Harmush discussed in detail how last summer, following the conflict with Israel, "Some of the roving bands took over private and public lands, and built houses there, as though these lands were spoils of war! So what Islam are they talking about?"

Since the end of November, participants in Sunni Islamist Web forums such as www.mohajroon.com have written that a battle between Sunnis and Shiites in Lebanon is imminent, and they have proposed steps to prepare for this war. Among the ideas discussed have been gathering intelligence on Shiite weapons arsenals in order to seize them, purchasing houses and warehouses throughout the country to serve as bases for the Sunni mujahedeen, and "liquidating" Shiite imams and leaders.

As the crowds of Hezbollah supporters fill the streets of Lebanon, it appears that any day now some action will spark further instability in the country.

Holocaust Deniers Anoint Iran's Ahmadinejad

December 12, 2006

> *"For 60 years, talking about the Holocaust was a crime in the West, but now there is a serious debate about the Holocaust in the media and also in the political and popular meetings."*
>
> — PRESIDENT AHMADINEJAD OF IRAN, DECEMBER 9, IRAN STATE TV

This week, the Islamic Republic of Iran is holding its third conference devoted to denying the Holocaust since President Ahmadinejad came to power.

Much more than just insulting the memory of 6 million lies behind why Holocaust denial has become state policy in Iran. Something more sinister is at play.

As an adviser to the Iranian president, Mohammad Ali Ramin, said, according to the Iranian reformist daily Rooz of June 9, "The resolution of the Holocaust issue will end in the destruction of Israel."

Yesterday and today, 67 "foreign intellectuals" from 30 countries are taking part in "Holocaust World Prospect," a conference sponsored by the Iranian Foreign Ministry's Institute for Political and International Studies.

According to a Mehr News Agency report from December 5, the focus of the conference is "whether the Holocaust really happened." Six official sessions are devoted to uncovering the "truth" about the Holocaust.

In his opening speech yesterday, Foreign Minister Manouchehr Mottaki said the conference was aimed at creating an appropriate scientific atmosphere for discussing the Holocaust. The host of the conference, Raoul Mousavi, added that Iran was providing an opportunity to discuss the Holocaust "away from Western taboos."

Attendees include a who's who of Western anti-Semites and Holocaust deniers. A former head of the Ku Klux Klan, David Duke, is giving the keynote address.

French professor Robert Faurisson told participants yesterday the Holocaust "never existed" and praised Mr. Ahmadinejad, who "will allow the truth to emerge in the end and uncover this great lie." Frederick Toben of Australia's Adelaide Institute told the conference that "claims ... about the mass gassings and burnings [are] a physical impossibility," while French writer Georges Thiel called the Holocaust "an enormous lie." Bradley Smith, of the America-based Committee for the Open Debate of the Holocaust, said, "Gas chambers never existed."

Of all the attendees, a group of anti-Zionist rabbis from America and Australia have received the greatest acclaim in the Iranian press.

Deputy Foreign Minister Manouchehr Mohammadi gave a series of news briefings leading up to the start of the conference yesterday. On December 6, he provided background for why the conference was occurring. The Islamic Republic News Agency quoted him as saying, "The points raised by President Ahmadinejad on the Holocaust a year ago resulted in a bizarre worldwide hue and cry. ... The president simply asked whether an event called the Holocaust had actually taken place." He also told the Fars News Agency on December 5 that "Iran, as an unbiased state, would accept the truthfulness of the Holocaust if sustaining evidence is presented."

This week's conference is just the latest chapter in Iran's Holocaust denial, which has risen to unprecedented levels since Mr. Ahmadinejad became president. Exactly one year ago this week, he stated live on Iran TV for the first time that the Holocaust was a "myth."

Many of the world's best-known anti-Semites praised Mr. Ahmadinejad's statement, as did leading Western Holocaust deniers — who now have an undisputed leader. A month after the president's speech, the Iranian Foreign Ministry announced that it would sponsor the first of multiple events that would examine the "truth" behind the Holocaust.

On February 14, the Iranian ambassador to Portugal, Mohammad Taheria, discussed a spring conference Iran was planning to assess the true scale of the Holocaust. "It is necessary that historians meet, discuss, and give their opinion, talk about numbers," Portugal's Public Radio Antena 1 and the Mehr News Agency quoted him as saying. "When I was ambassador to Poland, I visited Auschwitz and Birkenau twice and I did my own calculations. To incinerate 6 million people you would need 15 years."

In the first week of March, the Bassij, a branch of the Iranian Revolutionary Guard Corps, held its own conference, "The Holocaust: Myth or Reality," initiated by the Iranian supreme leader, Ayatollah Ali Khamenei, according to Islamic Republic of Iran Broadcasting.

Another conference came in the last week of June. According to IRIB, "Scholars and people gathered in Tehran to examine the case of the Holocaust at a distance from the world hegemony's media propaganda. ... The conference ... aimed to open the door to a fair judgment of the widespread claims about the mass killing of Jews during World War II."

Next up for the Iranian government's event planners is a conference "to study the genocide of the American aborigines by European immigrants" on February 27.

A Holocaust Denier's Victory

December 20, 2006

President Ahmadinejad of Iran was absolutely right to boast that "now there is serious debate about the Holocaust in the media and also political and popular meetings."

The international press has paid rapt attention to every one of Mr. Ahmadinejad's statements on the Holocaust. He is given unprecedented space in newspapers, TV, and magazines such as Time, whose current edition includes a "second exclusive" interview in three months. As the New York Times reported on Thursday, now Mr. Ahmadinejad "has a platform to promote the theories — and try to position himself regionally as the reasonable man who is asking hard questions."

The Holocaust deniers who participated in the conference in Iran last week also have been given an unparalleled opportunity to express their views in prime time on Fox News, MSNBC, and CNN. The latter two aired lengthy interviews with a former Ku Klux Klan leader who was the keynote speaker at the conference, David Duke.

Press outlets have given considerable space to the anti-Zionist Neturei Karta rabbis, whose hateful views were discussed in a long Washington Post article on December 14. In the past year, this group of miscreants — mislabeled in the Iranian press as "chief" and "prominent" rabbis — has met with the leadership of the terrorist groups Hezbollah and Hamas.

According to Iran's Fars News Agency, at the conference the rabbis lauded Mr. Ahmadinejad's "friendship with the Jewish community." No mention was made that family members of the Neturei Karta leadership were murdered at Auschwitz.

Mr. Ahmadinejad has also succeeded in uniting the world's leading Holocaust deniers into one body and under one leadership. The establishment of an Iran-led "World Foundation for Holocaust Studies," charged with finding out "the truth about the Holocaust," was announced December 14.

The Iranian president instructed his Foreign Ministry and embassies throughout the world last week to push Holocaust denial. The Iranian ambassador to Indonesia, Behrouz Kamalvandi, gave an interview to Indonesia Metro TV on December 13. In response to a question about the outcome of the conference, Mr. Kamalvandi said the "first and most evident result of the event was to bring up the issue in various countries." And the Iranian ambassador to Syria, Mohammad Hassan Akhtari, told the Italian daily Corriere della Sera that the "Holocaust is false" and the "gas chambers and the Treblinka and Auschwitz extermination camps have never existed."

According to the Iranian press, many of the Holocaust deniers who participated in the conference now worship Mr. Ahmadinejad. One Swiss researcher was quoted by Fars News on December 12 as saying the Iranian president is a "God given gift who has come to help to the destruction of the myth of the Holocaust."

"French Scholar Views Ahmadinejad's Words as a Miracle," was another Fars News headline on December 12. The article quoted an unnamed French attendee as saying conducting research or speaking about the Holocaust had not been possible for the past 20 years, and "the miraculous question" that Mr. Ahmadinejad "raised recently has now created a fully new condition regarding this issue." The French scholar added that the "Israeli state must be annihilated."

The conference has yielded some surprising results. Although the Islamic Republic News Agency quoted Mr. Ahmadinejad as saying the participants "enjoyed the taste of freedom" while in Iran and that once they "return to their homelands, their governments will not bother them," he was mistaken.

The French Holocaust denier Robert Faurisson — who said at the conference, "I declare that the chimneys you claim Hitler had built to burn the Jews is a sheer lie" — is now under investigation by the French authorities.

A Swedish high school teacher, Ian Bernhoff, claimed at the conference that only 300,000 Jews were killed by the Nazis. When he returned home, he was fired.

Thirty-four international NGOs that previously had relations with the Iranian Foreign Ministry's Institute for Political and International Audiences, which organized the conference, have now cut all ties to it.

In response, Foreign Minister Manouchehr Mottaki of Iran accused the West of mistreating the Holocaust deniers: "They try to charge them with different allegations and thus block the path of research on Holocaust on every excuse."

For the most part there have been strong reactions in the West against Iran in the week since the conference, including a group of Russian duma deputies canceling a visit yesterday to Tehran. But there has been very little reaction from Iran's Arab neighbors.

A Somali-born former Dutch member of parliament, Ayaan Hirsi Ali, eloquently wrote in the International Herald Tribune on December 15: "I cannot help but wonder: Why is there no counter-conference in Riyadh, Cairo, Lahore, Khartoum or Jakarta condemning Ahmadinejad? Why is the Organization of the Islamic Conference silent on this?"

Conspiracy Theories Bloom In Arab and Iranian Press

December 27, 2006

A program on Hezbollah's Al-Manar television network last week announced that Israel "murdered" a Lebanese parliament member, Pierre Gemayel, "with the help of "four American experts" who work at Beirut's airport. The announcement was quickly reported on by other Middle East press outlets, including Islamic Republic of Iran Broadcasting.

Conspiracy theories in the Arab and Iranian press regarding America, Israel, Jews, Christians, and Zionists have been extensively documented in this column. This year, the theories most cited focused on the attacks of September 11, 2001, the Holocaust, and the forces supposedly behind the strife in Iraq, Darfur, Lebanon, and elsewhere in the region.

With the advent of the Internet and satellite TV, conspiracy theories originating in the Middle East are now easily spread throughout the Muslim world — from Turkey to Malaysia.

In Iran, leading government and religious figures, who have absolute control of the press, have been behind some particularly nasty and outrageous conspiracy theories. The day before the fifth anniversary of September 11, the Iranian press was full of them.

The head of the Islamic Center in Washington, D.C., Muhammad Al-'Asi, was on Iranian TV saying that the "American administration" was behind the attack, while Lebanese Shiite leader Muhammad Hussein Fadhlallah blamed the Jews and some American powers.

Many other Iranian conspiracies are extremely anti-Semitic. An adviser to President Ahmadinejad, Mohammad Ali Ramin, was quoted in a daily newspaper saying Jewish treachery was behind such ills as the plague, typhoid, AIDS, SARS, and bird flu.

Conspiracies about "the Zionist regime" are also rampant in the Iranian press. IRIB reported that more than 10,000 Russian women are kidnapped each year and taken to Israel through Egypt by Zionists, and that each woman is sold for between $4,000 and $10,000.

The Iranian news channel IRINN asserted that Disney's "Pirates of the Caribbean "is a "Zionist conspiracy" by the company to exert cultural influence. Similarly, IRINN ran a special about "Zionist companies" such as Coca-Cola, McDonald's, and Pepsi, which it claimed stands for "Pay Each Penny Save Israel."

The secretary of Iran's Supreme National Security Council, Ali Larijani, was quoted by the Islamic Republic News Agency regarding reports on the "formation of terrorist squads in Brussels by Israel and certain European states for assassination" of Iranian dignitaries.

Official Friday sermons are used by religious leaders in Iran to smear Jews, Christians, America, and Israel. On May 19, the Ayatollah Kashani was quoted by the Tehran Times blaming "the CIA" and "Zionists" for terrorist attacks within Iran, and in particular targets connected to its economy, universities, and science.

The Iranians also try to influence the Muslim world to oppose American activity in Iraq, and they frequently make up things to tarnish America's image. For example, IRIB reported on April 5 that "U.S. troops are distributing boobytrapped dolls among Iraqi children."

An Iranian Foreign Ministry spokesman, Hamid Reza, was quoted by the Iranian Student News Agency on September 6 stating that terrorist attacks throughout the Middle East "are being guided by the White House."

IRNA reported on November 27 that the "apparent discord among Shiites and Sunnis in Iraq, Pakistan, and Afghanistan is the fruit of a conspiracy by America and the Zionist regime."

The Iranian daily Keyhan claimed on November 15 that Saddam Hussein was sentenced to death because he knows many secrets about American activity in the Middle East and that "the neoconservatives have to silence him once and for all."

Conspiracies about alleged Israeli activity in Iraq are also common, such as an article titled "Mossad Assassins Targeting Iraqi Scientists" in the Tehran Times, which reported that "more than 2,500 Iraqi scientists over the past 3 years were killed by the Mossad."

Another report from IRIB said, "Zionists will be burying nuclear waste in Iraq," which "is a threat to ... the people living in Iraq, Syria, and Jordan." The report claimed American forces in August burned poisonous substances in Fallujah.

Next week's column will be devoted to the most disturbing conspiracies from the Arab and Turkish press.

2007

Middle East Conspiracy Theories

January 4, 2007

> *"I think if the U.S. did not exist some of us would have invented it. ... It is because we are used to hanging all our problems and catastrophes on America. ... We add Israel to America."*
>
> — A FORMER KUWAITI OIL MINISTER, ALI BAGHLI, KUWAITI DAILY AL-SEYASSAH, MARCH 30

On her winter vacation to Saudi Arabia in December, a professor at Brandeis University whose work has been promoted by the Saudi Embassy in Washington, D.C., Natana DeLong-Bas, gave an interview to Asharq Al-Awsat.

"I do not find any evidence that makes me agree that Osama bin Laden was behind the attack on the twin trade towers," she said of the terrorist attacks of September 11, 2001.

To any observer of the Middle East, such a statement comes as no surprise, even from a university professor.

In an article for the September 10 edition of the New Sunday Times, "Did the U.S. Stage a Lie on 9/11?" the vice chancellor of the University Sains Malaysia, Dato Dzulkifli Abdul Razak, also questioned the official version of the attacks. And the Malaysian cleric Hussain Ye appeared on Peace TV on November 1 and said there was no proof Muslims were involved in the attacks and that Jews are guilty.

An article in the November 22 edition of the Syrian government-controlled newspaper Tishreen said a former secretary of state, Henry Kissinger, was connected to the September 11 attacks.

The article also criticized those who blamed Syria for the assassination of a Lebanese government minister, Pierre Gemayel. The Syrian minister of expatriates, Bouthaina Shaaban, instead blamed American Embassy employees in Beirut, as well as Israel, for the Gemayel assassination in the October 12 issue of Asharq Al-Awsat.

Conspiracies surrounding Darfur also abounded in 2006. In an address before the U.N. General Assembly on September 19, President al-Bashir of Sudan said that what was really happening in Darfur was a Zionist plot to dismember Sudan and plunder its resources.

The "American-Zionist interest in Sudan" is not to prevent genocide in Darfur but to get control of oil and uranium, Muhammad Salahuddin wrote in the Saudi daily Arab News on August 10. And "American-Israeli" "fabricated lies" about Darfur are part of a Zionist conspiracy to control the "Nile basin to the Euphrates River," the Sudanese writer Muhammad Keshk wrote in the Syrian government-controlled daily Al-Thawrah on December 14, while America is "encouraging the Christians of south Sudan to break away from Sudan," Hassan Tahsin wrote in the Arab News of June 23.

Anti-Semitic conspiracies also continued unabated in 2006 in the Arab press. In the Iraqi magazine Al-Shabaka Al-Iraqiyya of March 13, the article "Look for the Jews" blamed Jews for the cartoons of Prophet Muhammad published in Europe and for the destruction of the Buddha statues in Afghanistan in 2001 and the Samarra mosque in February.

The Egyptian cleric Hazem Sallah Abu Ismail, a former Islamic lecturer in America, appeared on Saudi Al-Risala TV on April 14 and discussed U.N. documents that purportedly showed that "82% of all attempts to corrupt humanity originate from the Jews." Six weeks later, Uwe Frisecke of Lyndon LaRouche's Executive Intelligence Review said on Lebanon's New TV that Jews spread AIDS, SARS, mad cow, and other diseases. The children's Web site of Egypt's Muslim Brotherhood devoted a section of its April issue to "murdering children — part of the Jewish religion."

Conspiracies about Guantanamo Bay were also common last year. Following the suicide of a Saudi man at Guantanamo, the man's father told the Arab News on June 19 that Zionists and neoconservatives in the American administration had masterminded his death.

"In countries and cultures where governments and the media have regularly colluded to hide the truth from their citizens, mistrust of authority is pervasive," the British foreign and Commonwealth office minister for the Middle East, Kim Howells, wrote of the Arab press in the London Arabic daily Al-Hayat on October 19. As 2007 begins, one can only hope that the conspiracy theories from the Arab press will lessen. This, however, is unlikely.

The Middle East Press Cheers the Rise of Russia

January 12, 2007

Many articles in the Middle Eastern press have been discussing Russia's growing influence and activities in the region, which are in direct opposition to American policy and are playing a negative role in the war on terror. To date, it is unclear what the American government is doing about it. One thing is for certain: Russia has big plans for itself in the Middle East.

Earlier this week, an unnamed Russian Defense Ministry official was quoted by Itar-Tass detailing Russian contracts of anti-aircraft weapon systems to Syria and Iran — worth more than $1.3 billion — are being filled on schedule. This follows Russia's continual support of Iran's nuclear program, in addition to its activities last summer instigating havoc in Lebanon, such as reportedly allowing Syria to pass Russian-made rockets to Hezbollah.

Leaders of Hamas, Egypt, Syria, and others have been establishing warm relations with Russia during recent visits to Moscow. President Putin plans to reciprocate with a trip to Qatar, Kuwait, and Saudi Arabia next month to sign lucrative "deals."

The November 2 issue of the Lebanese Daily Star noted that a visit to Moscow by President Mubarak of Egypt in November — to increase defense ties and sign large arms deals — was intended to make a point to America. An American expert on military studies at the Al-Ahram Center for Political and Strategic Studies, Mohammed Kadry Saeed, was quoted as saying that "this is a message to the American Congress" and that "the statements coming out of Russia regarding elevated military cooperation are a new development." Hanna Ebeid from the International Relations Unit of Al-Ahram Center for Political and Strategic Studies added that Russia could emerge as a "counterbalance" in the region.

The Arab and Iranian press have been rooting for the return and rise of Russia in the Middle East.

"Russia Stages a Comeback Through Syria" was the title of an article in the Daily Star of August 28, which reported, "The year 2006 has become the year of the Middle East for the Russian leadership. First was the controversial visit by a Hamas delegation to Moscow. … In the early summer, reports surfaced that Russia was engaged in modernizing facilities in the Syrian ports. … Then came the crisis in Lebanon. … The Russians have taken it upon themselves to be the international advocates for the Lebanese and Syrian governments. ... It is now beyond reasonable doubt that the Kremlin has decided to take Syria under its wing and use it to stage a 'comeback' to Middle East politics."

The article explained what was influencing Moscow's activities: "The rationale for Russia's new course in the Middle East lies in the same motivation that drives Moscow 's foreign policy as a whole: primarily, deep dislike of the United States combined with a desire to at least partly avenge Russia's defeat in the Cold War. … The idea that U.S. influence has to be curbed as much as possible and wherever possible is very popular among influential people in Putin's administration."

An editorial titled "Moscow Maneuvering to Regain Former Status" in the December 21 issue of the Tehran Times discussed the hopes of Russia restoring "the international status Moscow enjoyed during the Cold War" and added, "Amid tensions in the Middle East, Russia is trying to regain its former role in the region. … The main goal of these activities is restoring Russia's international status, which was greatly diminished after the collapse of the Soviet Union. … If Moscow can pull that off, Russia can establish a balance of power in the strategic Middle East region to counter U.S. influence."

The Syrian government-controlled Teshreen on December 20 included an article titled "The Russia Role" which stated: "Russia has started to restore its central and important role on the international scene in general and on the Middle Eastern scene in particular." According to the article, Russia "is on the rise."

A very troubling sign of things to come is that Russia's stated foreign policy being driven by its "deep dislike of the U.S." is being applauded by state-sponsored terrorist regimes in Iran and Syria and by allies such as Egypt and Saudi Arabia.

Arabic TV Network Broadcasts 'Jihadi Porn'

January 24, 2007

Following in the footsteps of Al-Jazeera and Al-Manar, the Arabic-language TV network Al-Zawraa is grabbing headlines in the West.

The station was founded by a Sunni former member of the Iraqi parliament, Mishaan al-Jabouri, who was indicted in February 2006 for allegedly embezzling millions of dollars designated to protect oil pipelines in Kirkuk but funneled instead to Sunni insurgents.

The channel, whose name means First Channel in Arabic, started out broadcasting music videos and other popular culture shows, but over the past two months it has become an anti-American propaganda outfit. Its programs are vehemently anti-Shiite.

Al-Zawraa — which has been misnamed as "Al-Qaeda TV" by press outlets such as Fox News and the Weekly Standard — airs 24 hours a day on Nilesat, reportedly from a moving vehicle in Anbar province in western Iraq. Although Nilesat is under the control of the Egyptian government, the Egyptian information minister, Anas al-Faiki, has been quoted in the Arab press as saying, "We have no concern what the channel is doing."

The best way to describe Al-Zawraa is as a channel devoted to airing the "jihadi porn" that can be found on every Islamist Web site. This includes music videos of terrorist attacks against American forces, with Islamic chanting heard in the background.

Many segments feature masked men shouting "Allahu Akbar" as they detonate bombs. One clip from last month included a long speech cursing President Bush, while another showed what was identified as the shelling of American troops at the Baghdad airport.

The infamous jihadi porn video "Juba: Baghdad Sniper," which was produced by the organization Al-Jaysh Al-Islami Al-Iraqi and presented to viewers of Islamist Web sites as a holiday gift to believers in Allah's unity ("ahl al-tawhid"), has been shown multiple times on Al-Zawraa.

The video opens with a masked sniper shooting someone and then entering a room and adding another name — no. 37 — to a hit list in his notebook. An announcer reads aloud what the sniper is writing, with simultaneous English translation. The speaker calls upon the Islamic nation to expel from Muslim lands the Crusaders and Jews, whom he accuses of destroying mosques, attacking the Koran, and violating the honor of Muslim women. He repeats a Koranic warning: "If you do not go out" to jihad, Allah "will punish you heavily and will replace you with other people."

Throughout the video, the speaker emphasizes the importance of videotaping killings, and he shows footage of American soldiers being shot by snipers. At the end of the video, a chart appears, showing that 668 American soldiers have been killed by sniper fire over the last year, from Ramadan to Ramadan. In another program, a Zawraa host criticizes the Iraq Study Group report, calling it "naked" interference and saying it will take Iraq back to "point zero."

A State Department official told Voice of America on January 18 that the American Embassy in Cairo has "raised concerns about Al-Zawraa" with Egyptian officials and asked that the station be removed from Nilesat.

But the chief of Nilesat, Amin Bassiouni, has said that as far as he is concerned, it is simply a business decision to air Al-Zawraa, and that Nilesat must honor its contract with the station's owners. He said that neither the Iraqi nor that Egyptian government has requested that the channel be shut down.

According to Mr. Jabouri, Al-Zawraa is gaining viewers among Sunni Muslims.

Iran's Economic Crisis

January 31, 2007

Iran's civil society is experiencing major breakdowns, the country's reformist press and Web loggers are reporting. Signs of growing economic instability include high inflation, rising prices, food shortages, and long lines at gas stations.

In the first week of January, 150 deputies at the Majlis, Iran's Parliament, openly criticized President Ahmadinejad's handling of such issues. The same week, a group of 54 influential Iranian professors wrote an open letter to the government about its agricultural policies.

The professors discussed Iran's "rapidly depleted" natural resources, such as water, and criticized the country's growing reliance on imported grains and rice, as well as the deterioration of its farms and ranches. They said many Iranians, in particular the young, elderly, and poor, have begun to "suffer from severe nutritional deficiencies, which threaten Iran's public nutritional security."

The January 20 edition of the Iranian reformist daily Rooz included an interview with an economist, Saeed Leilaz. "Iran is on the verge of economic collapse," he said. "A large portion of the economic turn for the worse is due to Ahmadinejad's policies and management style ... [which] have prompted many to publicly criticize [him]. ... The administration has increased government expenditures so much that we will face an enormous budget deficit in the coming year."

Mr. Leilaz also said Iran's deep economic crisis could "ultimately lead to the disintegration of the government."

Gas shortages throughout the country, as well as news of Iranians "waiting in lines for hours to purchase gas cylinders to heat their homes or cook food," were detailed in a report in Rooz on January 14. In some towns, lines of people were three miles long, and citizens were upset about the rising prices of produce, dairy, rice, eggs, cereal, and grains, the paper reported.

The Arab press has been watching Iran's economic crisis closely. An editorial in the Saudi daily Arab News of January 24, "Frustrations in Iran," focused on Tehran's economic travails and "the growing criticism of Ahmadinejad's presidential performance."

"The problem for the president is that the economy is weak," the editorial said. "Despite its oil wealth, life is hard for ordinary Iranians. ... There is therefore a groundswell of frustration because of the president's economic neglect."

A Saudi columnist, Ahmed Al-Rabei, wrote in the January 29 issue of Asharq Al-Awsat about the Iranian economic crisis. "Instead of attempting to solve the unemployment and poverty problems of its people, the new revolutionaries decided instead to 'export the revolution,'" he wrote. "After all, a revolution is not a commodity to import but rather the creation of a new regime that sets an example to others. What the revolutionaries did in Tehran was a disaster rather than an example."

Mr. Rabei criticized Iran for spending money on spreading war rather than helping its needy. "Iran is actively involved in Lebanon," he wrote. "They are the real motivators of war and chaos through the 'clean money'" — Iran's funding of Hezbollah — "that they appropriated from their poor. They are also involved in both Iraqi and Palestinian affairs. They are involved in acts of violence in Yemen and are nauseatingly active in Syria, Egypt, Pakistan and other nations while at home their unemployment and poverty problems worsen and their oil revenues are channeled into the development of nuclear weapons technology and the building of its military arsenal."

American policy-makers should take advantage of Iran's dire economic situation. Targeted financial measures could lead to the ousting of Mr. Ahmadinejad, who is walking and talking like the leader of a superpower and openly challenging America at every step.

His most recent provocation, creating a $2 billion social investment fund with Venezuela this month "to challenge America," is not going over well with the average Iranian, who is waiting in long lines for food and gas.

Many in Iran would be happy to see a regime change. The American government needs to reach out to these people.

Iran's Top Strategist, In His Own Words

February 14, 2007

Western policy-makers are all familiar with the names Khomeini, Khamenei, Khatemi, Rafsanjani, and now Ahmadinejad. But what about Yahya Rahim Safavi? Chances are, most do not recognize his name — but they should.

Yahya Rahim Safavi is the major general in command of the Islamic Revolutionary Guards Corps of Iran. He has been a force in the Iranian military since he rose to fame as a young leader in the Iraq-Iran war of 1980–88.

In December, General Safavi's name was included in U.N. Security Council resolution 1737, which called for his assets to be frozen because of his involvement in Iran's nuclear program.

The general has often made outrageous public statements about America. In September, he told the Iranian Students News Agency that America's current policy "of pre-emption and war mongering in order to dominate the Middle East" is part of a plan initiated during "the events of September 11, 2001," which "were ordered by the U.S. and Mossad."

Even though General Safavi is one of the most important military leaders in Iran, his statements are virtually ignored by most press outlets. American officials, however, should take them seriously.

During a press conference at the Revolutionary Guards' naval forces headquarters on February 3, General Safavi said: "In the event of any kind of military threat ... Iran's armed forces will provide a crushing and inconceivable response against the invaders, and will endanger the interests of the foreigners in the region. ... [It] will bring in its wake American humiliation in the world."

Although the general disclosed some important information about Iran's military doctrine, he added: "We never reveal all our cards to the enemy. ... We will respond to any invading power with a force that they cannot even imagine. Neither the Americans nor the Zionists know what complex, precise, and intelligence-based plans we have designed."

In a November 12 interview with Iran's Channel 2, Commander Safavi discussed his country's military strategy vis-à-vis America: "The Americans have many weaknesses. … In the wars in Iraq and Afghanistan, they clearly displayed" them. … "We have planned our strategy precisely" around them. … "We don't see any motivation among the American forces in Iraq. They are very cowardly."

One of the lasting memories of the Iraq-Iran war is the thousands of Iranian children sent on suicide missions to search for land mines planted along the border. The "martyrdom" missions are a key component of General Safavi's military planning.

In an interview on Al-Alam TV on November 5, he said: "We naturally have defense plans of deterrence. ... The IRGC has thousands of martyrdom-seekers, who have gained military experience during the eight years of war, and these martyrdom-seekers are prepared to carry out martyrdom operations on large scales. They operate professionally and have undergone training. They have a strong spirit of martyrdom."

On October 28, 2005, he had said on Jaam-e Jam3 TV: "The Americans and Israelis should know that the future belongs to Islam. … The Muslim peoples in Iraq, Afghanistan, and Palestine will eventually cut off the legs of the American, Zionist, and English occupiers from the Islamic lands."

To mark Basij Day on November 21, General Safavi boasted, "The Americans are sunk in the quagmire of Afghanistan and Iraq, and there is no way for them" to move either "forward or backward. Assuming they attack Iran, their 200,000 troops, in their 33 bases, are highly vulnerable. … They can start a war, but" the decision to end the war "will not be in their hands."

Earlier last year, in an interview with the Iranian News Channel on March 29, General Safavi described Iran's monitoring of American troop activity: "If they turn to military threats, they should know that the armed forces of the Islamic Republic are fully prepared."

On August 27, in an interview with Iran's Channel 2, General Safavi described an incident that he claimed had occurred between American and Iranian forces. "A few months ago, the USS Ronald Reagan aircraft carrier … which was carrying on deck 72 F-18 planes, and which was escorted by several warships, was circled for about 25 minutes by one of our unmanned planes, which filmed and transmitted the footage. … We have shown them that they are very vulnerable. Even their aircraft carriers and other warships are vulnerable," he said.

The Iranian Military Targets American Forces

February 21, 2007

> *"We say to America that if it ignites the fire of war, it will doubtless engulf the White House more than it burns others."*
>
> — THE IRANIAN DEFENSE MINISTER, MUSTAFA NAJAR, ON AL-JAZEERA TV, FEBRUARY 6

The head of the ground forces of Iran's Revolutionary Guards, Nur Ali Shushkari, told Iranian government news agencies earlier this week that one of its commando units had engraved its emblem into the side panel of an American warship stationed in the Persian Gulf. The point of Mr. Shushkari's bold statement was to make it clear that the Iranian military is watching every move made by American troops in the region.

Other leading Iranian military, political, and official news outlets have made similar statements. On November 27, the Mehr news agency quoted the navy commander of the Islamic Revolutionary Guards Corps, Admiral Sejjad Kouchaki, as saying, "We are fully monitoring the route taken by the American" warships in the Gulf, "and because American warships are heavy, they have no maneuverability and are easily sunk."

During war games in mid-August, Admiral Kouchaki gave a press conference, covered by IRINN TV, during which he said Iranian submarines "can shadow the enemy's vessels very closely" and at which he described Iran's naval tactics vis-à-vis American forces.

Admiral Kouchaki said, "Our submarines can easily get near the enemy. Even our enemies know full well that one of our submarines passed under one of their [vessels] without their noticing. We came close to their anchored vessels, and we even filmed their anchor chain. We followed them through a periscope at a depth of 1 kilometer without their noticing. All these methods and tactics are nonclassical and nonconventional, constituting unbalanced warfare."

At the same press conference, the head of the Iranian army's joint staff, Abdolrahim Mousavi, added, "Regarding what Admiral Kouchaki said ... The same goes for unmanned planes. The same unmanned plane that flew over the warships in the Persian Gulf, taking pictures of them all, can do other things if equipped with a warhead. No matter how large they are, they have vulnerabilities."

"Is War on the Way?" was the title of a January 22 editorial in the Iranian daily Kayhan, which noted: "The Americans must be warned ... The American soldiers to the East and West of the Iranian" border "are in range of our fire. [When] the mighty missiles are launched from Iran, Israel will become a scorching hell for the Zionists, [in which they will burn] before reaching actual hell."

Sobh-e Sadeq, who is the mouthpiece of the Revolutionary Guards, reported in a January 29 article that elements in the Revolutionary Guards threatened to abduct American soldiers around the world. "When the American security officers fall easily to the untrained and inexperienced Mahdi army forces ... and when a ship belonging to the English forces disappears ..." in the Persian Gulf "under unclear circumstances — all these can constitute a clear message."

The article further stated, "All it would take is for [Iran] to open its wallet a little" to its supporters in various parts of the world, "and we will witness long lines of blond, blue-eyed officers who will become the prisoners of the fighting cocks who wait only for a signal — and a word to the wise is sufficient."

In a February 10 television address on Iran Channel 1, the secretary of the Iranian Guardian Council, Ayatollah Ahmad Jannati, warned, "The entire world hates" America. "How will they dare show up here? Once they were beyond our firing range. Today, the Americans are within our firing range. They have placed us under siege. But, in a way, this is to our benefit because they are within our firing range now. In the West, in the East — they are within our firing range everywhere."

The deputy head of the Revolutionary Guards, General Zol-Qadr gave an interview on Islamic Republic of Iran Broadcasting TV, during which he explained that by monitoring American forces, Iran had "a great opportunity ... to get to know ... American methods, strategies, doctrines, and tactics."

He added: "We have been able to prepare for a confrontation with the American threat. Today, thank God, we can say the armed forces in our country are totally familiar with the American modus operandi. ... We even know the American commanders' state of mind. ... This increases our maneuvering ability against the American threat."

This interview occurred in February 2005. Clearly, the Iranian military has been in the planning stages for years for a possible confrontation against America.

Iran's Ahmadinejad And the Darfur Deniers

March 7, 2007

> *"That Sudanese public radio is still calling Israel 'the enemy' ... is a good reason why the ... Zionist groups all over the world, especially in America, are dedicating so much attention to the issue of Darfur when it is not the only humanitarian crisis in the world."*
>
> — THE SECRETARY-GENERAL OF NORTHERN SUDAN'S RULING NATIONAL CONGRESS PARTY, IBRAHIM AHMED OMAR, FEBRUARY 2007

Here are the latest facts about Darfur. In three years, more than 200,000 have been killed — some estimate 450,000 — and more than 2.5 million displaced.

On February 27, the International Criminal Court announced that it planned to try the Sudanese minister for humanitarian affairs, Ahmed Mohammed Haroun, and a janjaweed leader, Ali Kushayb, for crimes against humanity. The Sudanese interior minister, Al-Zubair Bashir Taha, responded by telling the Arab daily newspaper Al-Hayat on March 1 that the Sudanese government would "behead" any person who tried to arrest Sudanese officials.

At the same time, some are downplaying and even denying the killings, rapes, and displacements taking place in Darfur. The deniers are a diverse group, ranging from the Sudanese and Iranian leadership to the Arab government-controlled press, which would all rather blame the usual suspects — Israel, the Zionists, and America — than admit what is happening there.

President Ahmadinejad of Iran is at the top of this list. On Thursday he met in Sudan with the Sudanese leadership. Reports from the Iranian press quoted him as expressing pleasure at the "tranquility" of Sudan.

Referring to Darfur, he said, "There is no place in the world that suffers from divisions and wars unless America or the Zionists' fingerprints are seen there." According to the Islamic Republic of Iran Broadcasting news network, Mr. Ahmadinejad urged Islamic states to thwart such "conspiracies."

His statement followed a series of meetings between Iran and Sudan. On

January 15, the Sudanese defense minister, Abdel Rahim Mohammed Hussein, visited Tehran and tried to duck responsibility for what is happening in Darfur, claming there really is a "satanic plot" by "the U.S. and Israeli regimes, [which] are working hard to incite the conflict."

President Bashir of Sudan told attendees at the Nation of Islam's national conference in Detroit via satellite on February 24 that America is "exaggerating troubles in Darfur" so it can control the country as it has Iraq. His comments were broadcast live on Sudanese state TV.

"Bashir Reveals Zionist Plot in Darfur" was the headline of a September 20 story on IRIB that included Mr. Bashir lambasting "a Zionist plot to dismember his country and plunder its resources … particularly its oil reserves and then place it under a de facto U.N. trusteeship. … He said humanitarian agencies are exaggerating the extent of the suffering of Darfur civilians to secure increased funding."

In the February 22–28 edition of the Egyptian government weekly Al-Ahram, the secretary-general of northern Sudan's ruling National Congress Party, Ibrahim Ahmed Omar, was quoted as saying: "The West wants to see Sudan divided. … We know it is not just about Darfur. Such a division of Sudan … would serve many of the West's interests, and for that matter of Israel, the Western agent in this region."

Mr. Omar added that the West is seeking to control Sudan's natural resources and that "the Americans cannot accept the fact that Sudan has large and very much unexploited oil reserves while it is not bowing to the will of Washington."

In a cartoon in the February 8-14 edition of Al-Ahram Weekly, a blind-folded Sudanese speaker is shown. Behind his back is Uncle Sam with butcher knives cutting up Sudan, which hangs on a meat hook. Under Uncle Sam are two flagged dogs that represent Israel and Britain, licking their lips as Sudan bleeds.

"The Real Conspiracy," an article by Ayman El-Amir, is laid out under the cartoon. "It is safe to assume conspiracy is at work" in all conflicts in the Middle East that are perpetrated by Israel, Mr. Amir writes. "Westerners have often scoffed at Arabs as conspiracy theory addicts. Throughout the 20th century, everything Arab nationalists suspected as a scheme by colonial powers against their interests and aspirations was dismissed as a figment of Arab imagination. … Yet after decades of secrecy, declassified documents of confidential meetings, agreements, diplomatic correspondence, and reports, from the early years of the last century to the mid-1990s, reveal that the stretch of Arab imagination is much narrower than the scope of the conspiracy."

Sadly, Darfur is not so far from Cairo, where these words were written. As the killing continues, many in the Arab world and Iran look the other way, and instead of helping, they create conspiracies to justify their inaction.

Al-Jazeera's Dirty Little Secret

March 14, 2007

"Al-Jazeera Satellite Network, the Greatest Arab Media Organization."

— IKHWANWEB.COM, THE WEB SITE OF THE MUSLIM BROTHERHOOD, OCTOBER 25

Since the war on terror began, many Arab reformists have spoken out against "the culture of hate" that has incited its people to terrorism. The TV network Al-Jazeera is often at the center of this criticism.

Arab reformists who have witnessed first-hand incitement on Al-Jazeera often discuss its connection to the Ikhwan movement, aka the Muslim Brotherhood. This organization is one of the world's leading Islamist groups, based in Egypt and founded in 1928 by Hasan al-Banna. Today, its ideology influences groups ranging from Hamas to Al-Qaeda.

One of the most prolific Arab scholars and columnists, Mamoun Fandy, wrote in the London-based Arabic daily Asharq Al-Awsat about the movement's relationship with Al-Jazeera on January 22. "The Muslim Brotherhood has at its disposal media that transcends borders, from newspapers to satellite channels, which have taken over the minds of millions ... throughout the entire Arab world," he wrote.

Calling Al-Jazeera "the Muslim Brotherhood channel," Mr. Fandy discussed the network's activity in the Middle East. "Al-Jazeera ... is already propagandizing for the organization. ... If you watch a debate program presented on Al-Jazeera ... you will be amazed at the supreme effort ... to defend the Muslim Brotherhood."

Pierre Akel, founder and editor of a prominent Arabic reformist Web site, Metransparent.com, summed it up on February 9, 2006, "When it comes to satellite television in the region, Al-Jazeera is controlled by the Muslim Brotherhood."

A Tunisian intellectual and director of the Netherlands–based Center for Promoting Democracy in the Arab World, Khaled Shawkat, wrote about the connection between Al-Jazeera and the Muslim Brotherhood in an article for the progressive Arabic Web site Elaph.com on July 29. "Al-Jazeera has been hijacked by the Muslim Brotherhood organization," he wrote.

In an article in Metransparent on December 22, 2004, the Arab reformist Abu Khawla also discussed the Muslim Brotherhood's connection to Al-Jazeera. In addition, he wrote about the channel's religious authority, Sheikh Yusuf Al-Qaradawi, who as a member of the organization left Egypt 45 years ago, seeking asylum in Qatar. "Arabs turn nowadays to satellite TV channels, especially Al-Jazeera … a tool of communications that is so far completely monopolized by fundamentalists," Mr. Khawla wrote. "In the case of Al-Jazeera, preacher Al-Qaradawi (a leading figure of Egypt's Muslim Brotherhood) is one of its most influential founding members. And fundamentalists are in charge of its news programs and talk shows."

The always eloquent Youssef Ibrahim wrote about Sheikh Qaradawi and Al-Jazeera in his New York Sun column on Monday: "One of the most eminent leaders of the Ikhwan movement, who appears weekly on Al-Jazeera's 'Sharia and Life' program, is Sheikh Yusuf Al-Qaradawi. He has specifically ruled that Americans in Iraq and Israelis everywhere should be targeted by suicide bombers, who will be considered martyrs and heroes." Mr. Ibrahim also noted that Sheikh Qaradawi "ranks among the top 10 leaders of the Ikhwan's international ruling councils."

One Arab reformist critical of Al-Jazeera called it "the satellite channel of the masked terrorists," which "makes great efforts day and night to encourage terrorism," according to a September 5, 2004, article on Elaph.com by Abd Al-Khaliq Hussein. The article was highly critical of the channel and its connection to the Muslim Brotherhood, as well as Sheikh Qaradawi's position there.

Many leaders of the Muslim Brotherhood use Al-Jazeera for their own platform. The supreme guide of the Muslim Brotherhood, Mahdi Akef, appeared on Al-Jazeera on July 30, 2006, and said: "This American Satan claims to be a messenger of divine guidance. ... I salute all the honorable Americans who stand up to this Satan, who wants to set the entire world on fire, not just the Arab and Islamic nation. I go back to the issue of jihad. Jihad is an individual duty incumbent upon every Muslim."

A member of the General Guidance Council of the Muslim Brotherhood, Abdul Al-Munim Abdul Al-Futuh, appeared on Al-Jazeera on April 25, 2004, and said: "We support armed resistance. We, as the Muslim Brotherhood and as the world leadership of the Muslim Brotherhood, support the armed resistance in Iraq against the American occupier, 100%. I personally, like all members of the Muslim Brotherhood in Egypt and around the world, wish to carry weapons against the American occupation in Iraq." (More examples of this are online at memritv.org.)

Much confusion and controversy surround Al-Jazeera. To understand this network better, we should listen to what Arab reformists who are battling the culture of hate within their society are saying about it.

Iran Goes Berserk Over '300'

March 23, 2007

> *"The film is an insult to the rich Iranian culture ... it is surprising that Hollywood ... has begun to propagate hatred and terror in the world."*
>
> — PRESS RELEASE FROM THE IRANIAN EMBASSY IN PARIS, MARCH 22

Over the past few years, the Iranian government has produced films, TV movies, and music videos about Israelis stealing the eyeballs of Palestinian Arab children, the Statue of Liberty as Satan, and President Bush as the reincarnation of Hitler, among countless others in the same vein.

Hollywood is another popular target of Iranian state-controlled films. A recent notable example was a special on the "Tom and Jerry" cartoons as a Zionist conspiracy.

The Iranian News Channel on July 27 did a report on Hollywood as a "weapon" of the Zionist lobby, with Disney being its leader. The report cited "Pirates of the Caribbean" as an example.

Other Iranian government programs about Hollywood have been devoted to the "reality" of the "Protocols of the Elders of Zion," as well as claims that "The Diary of Anne Frank," "Schindler's List," "Driving Miss Daisy," and "The Matrix" are Zionist propaganda. Steven Spielberg is a frequent subject, as are what the government calls the "ugly Jewish actress Barbra Streisand" and the "deceptive Woody Allen, who confuses people about the Jews."

Therefore, Iran's response to the recent Hollywood film "300" — an adaptation of Frank Miller's comic book account of the Battle of Thermopylae between the Greeks and the Persians, 480 years before the common era — was interesting, to say the least.

Iranian government-controlled press outlets have gone berserk over the movie, which grossed more than $100 million within eight days. A sequel is already in the works.

Iranian members of parliament condemned it. According to the Mehr News Agency of March 13, they sent a letter to the Iranian Foreign Ministry and Ministry of Culture and Islamic Guidance stressing the necessity for "measures" to prevent the screening of the movie throughout the world.

An Iranian government spokesman, Gholam-Hossein Elham, said "300" is part of an extensive cultural aggression against Iranian culture. Iran's representative to UNESCO demanded a denunciation of the movie by the United Nations Educational, Scientific, and Cultural Organization, the Organization of the Islamic Conference, the Non-Aligned Movement, and the Group of 77.

In an interview with the Fars News Agency on March 11, an art adviser to President Ahmadinejad of Iran, Javad Shamqadri, said: "Following the Islamic Revolution in Iran, Hollywood and cultural authorities in the U.S. initiated studies to figure out how to attack Iranian culture. ... Certainly, the recent movie is a product of such studies."

The Fars Art Bureau in southern Iran was quoted in the Mehr News Agency on March 13 as saying: "No doubt, it is hard for the hegemonic powers to see a firm, religious Iran that has pricked their bubbles."

According to a March 13 report by the Islamic Republic News Agency, "The movie fabricated history. ... No Greek king dared to stand up to the Persian Empire."

An editorial in Taliya News on March 9 called the movie an attack by the "U.S. propaganda machine ... against Iran, its culture, and civilization." The article reported that some Iranians have begun petitions against Warner Bros., which released the film.

Iran's largest circulation newspaper, Hamshahri, contended that "300" is "serving the policy of the U.S. leadership" and predicted it will "prompt a wave of protest in the world."

"Hollywood Declares War on Iranians" was a headline in the March 14 edition of the Ayende-No newspaper. The article reported: "In Iran, the movie hasn't opened, but ... bootleg DVDs are already available on the streets. Iranian state-run television has run several commentaries ... calling the film insulting and has brought on Iranian film directors to point out its historical inaccuracies."

One such commentary aired on IRINN TV on March 13 and purported to "shed more light on the conspiracy behind the screening of this anti-culture Hollywood film." It said: "The Zionist Warner company is also pursuing ... a propaganda front against [the] ancient and historical roots of Iranians. ... This film tries to ... increase the international political pressure on Iran."

The Iranian response to "300," while entertaining, shows that the Iranian government lacks self-confidence.

The Word Games That Hamas Plays

March 28, 2007

> *"I am confident that the siege will be partially broken, which will give our people an opportunity to be prepared for the forthcoming stage."*
>
> — KHALED MESHAAL, ASHARQ AL-AWSAT, FEBRUARY 2

Has Hamas moderated? Or are the Palestinian Arab organization's leaders following in the footsteps of Yasser Arafat, discussing peaceful intentions in talks with the West while declaring jihad in Arabic?

During a press conference in Tehran with President Ahmadinejad on March 4, the political leader of Hamas, Khaled Meshaal, said he supported the recently signed Mecca Accords for power sharing between Hamas and the Fatah Party of Arafat and the president of the Palestinian Authority, Mahmoud Abbas.

Mr. Meshaal also spoke in favor of other Palestinian Arab objectives, such as the establishment of a state with the 1967 borders and Jerusalem as its capital, and the so-called right of return.

Mr. Meshaal reiterated Hamas's position against official recognition of Israel, and he promised to continue the group's resistance. With a smile on his face, Mr. Ahmadinejad told the Hamas leader: "The Zionist entity is in the worst period in its history, and is headed toward crumbling. ... The divine victory ... will soon be revealed."

At an address at Al-Murabit Mosque in Damascus on February 3, 2006, Mr. Meshaal said publicly what some Palestinian Arabs have been talking about for some time with their press outlets: They will be patient in their battle against Israel. He promised, "Muslims will take over the world," and he explicitly said his organization's plan is to deceive Israel with semantics.

In his speech, the Hamas leader explained that his people are willing to continue fighting Israel even if it takes 1000 years for victory. Mr. Meshaal also said one aspect of Hamas's current strategy is to rely on such tools as using statements like "we love peace" or "we have given up the option of war," while still planning Israel's destruction.

Mr. Meshaal also promised: "Before Israel dies, it must be humiliated and degraded. ... We will make them lose their eyesight, we will make them lose their brains."

A Kuwaiti journalist and former education minister, Ahmed Al-Rabei, wrote an article in Asharq Al-Awsat on March 30 titled: "Hamas Between Two Languages." He chided the organization for playing games with its words, explaining, "The Hamas movement will save the Palestinian people a lot of time and effort if it ... change[s] its political language. What does this government program really mean?"

Mr. Rabei accused Hamas of not openly stating its true objectives with Israel. In "clear Arabic language, it means recognizing that there are two states, one of them Israel. It means negotiating the last solution with Israel. In plain Arabic, it means Hamas's recognition of Israel as a state and the political process as a procedure. ... The Hamas leadership has a chance to tell its bases the truth. ... Hamas has new obligations that it must not avoid. There is no exit other than to stop talking with two different languages."

Hamas's language games have confused many — including an Al-Qaeda deputy, Ayman Al-Zawahri, who recently accused the organization of getting soft with its jihad. In response, as reported in the Turkish Daily Zaman on March 13, the organization released a statement that read, "We will not betray promises we made to God to continue the path to jihad and resistance until the liberation of Palestine."

Besides Mr. Meshaal, many Hamas leaders have been discussing the coming "stage," or battle against Israel. A member of the Palestinian Legislative Council and columnist for the Hamas organ Al-Resala, Younes Al-Astal, called the Mecca Accords "Sulh Al-Hudaybiyya," alluding to a temporary peace agreement from the early days of Islamic history.

He wrote on February 15: "We are optimistic that the new Mecca agreement will be the key to many achievements, which will provide the internal security that is necessary to renew the siege on Khaybar [Israel], and to subjugate the Zionists to the demands of the resistance."

Another sign that Hamas is preparing for an additional "stage" against Israel has been its press outlets' support for suicide bombers. This topic will be explored further in next week's column.

Hamas TV Prepares For Intifada III

April 12, 2007

Some Hamas leaders have been speaking in Arabic of "a coming stage" against Israel. Many Middle East analysts are also discussing a possibility of an "Intifada III." Unlike the first two "uprisings," the Palestinian Arabs now have press and broadcast apparatuses at their disposal.

Leading up to the Palestinian Arab parliament elections in January 2006, and following the Arab satellite TV revolution led by Al-Jazeera, the terrorist organization Hamas launched its own TV network named after the Al-Aqsa mosque. Early on, most of the programming was devoted to Koranic readings and images of Islamic holy sites in Jerusalem.

Shortly thereafter, Hamas officials were quoted saying it would be emulating what Hezbollah did with its TV network Al-Manar and that it would also aim to spread the ideology of Hamas worldwide.

The director of the board of Al-Aqsa TV, Fathi Hamid, told Islam Online in January 2006, "It's only fitting that the Islamic movement, Hamas, should have a TV station where we can explain our hopes, our Islamic culture, and counter the widespread and incorrect stereotyping of struggle and resistance as terrorism. Ultimately, we hope Al-Aqsa TV will be a bridge between Hamas and the entire world, so we can have our own voice in the international media."

Programming developed to include calls by Hamas leaders for Palestinian Arab viewers to become suicide bombers, as well as segments in support of the killing of Israelis. One example includes an address at a rally led by Sheikh Fayyadh Al-Akhdhar, a Hamas leader in Nablus, which aired on Al-Aqsa TV on December 23.

The Hamas religious figure said, "The Hamas of the Revolution of the Mosques, of the War of the Knives, of the car bombs, of the great martyrdom-seekers, of the Al-Qassam missiles, of the Zionists' death tunnels, of the kidnapping of the soldiers and the settlers is the same Hamas of the government, of the legislative council, and of the ministries. ... It will continue to march along the path of the martyrs and to follow in their footsteps until the liberation of the Al-Aqsa Mosque from the filth of the aggressive oppressors."

Over time, Al-Aqsa TV began to air memorials to "martyrs" such as Hamas founder Sheikh Yassin. More recently, segments inciting children to martyrdom have begun being broadcast. On March 8, the channel aired an interview with the two kindergarten-age children of suicide bomber Rim Al-Riyashi, who told the channel, "Mama went to paradise for killing Jews."

The father of suicide bomber Tareq Hamid was ecstatic as he praised his son's martyrdom operation to the channel on March 6. He told the interviewer that when his son was about to perform the terrorist attack, he called his friends and told them, "I swear by Allah that I saw the black-eyed virgins of paradise on the hood of my car."

Mr. Hamid added, "Some people claim that all these young mujahedeen who blow themselves up are desperate people. … These claims are wrong. These are lies and clear Zionist propaganda. All the mujahedeen I came to know … are the finest of Palestinian society. … The only thing they need is to reach paradise by means of defense and martyrdom for the sake of Allah…"

It is worth mentioning that the official Hamas newspapers also have been praising suicide bombers. One deplorable instance was in the November 27 issue of the Hamas biweekly Al-Risala, which published an article by Hamas spokesman 'Abd Al-Latif Al-Qanu' drooling over Fatima Al-Najjar, the Palestinian Arab grandmother who blew herself up in November near Israeli soldiers in the Gaza Strip. In the article, Mr. Al-Qanu' also praises resistance as a means of liberating what he referred to as the Palestinian Arab territories "from the [Mediterranean] sea to the [Jordan] river," and he calls for suicide attacks in the heart of Israel.

Calling her attack "a turning point in the Palestinian people's jihad," the Hamas spokesman wrote, "This woman revolutionary acted like a man, put an explosive belt around her waist, and advanced toward a group of soldiers. … She became the first Palestinian grandmother … to have carried out martyrdom. In her heroic act, Umm Muhammad sent a clear message. … We will all be human timebombs that will explode in the heart of the criminal Zionist entity."

One must ask, "Why are no leading Arab political or religious leaders denouncing a culture that is cultivating a society made of suicide bombers — from the youngest children to the most elderly?"

Dancing With the Muslim Brotherhood

April 19, 2007

> *"We ask Allah that the elimination [of America] will be at the hands of our mujahideen brothers, [and] that Allah will give them martyrdom for His sake, [and] will bring victory to Islam."*
>
> — MUSLIM BROTHERHOOD CHILDREN'S AFFILIATE, AWLADNAA. NET, APRIL 2006

During the past few weeks, a diplomatic mess has spilled out into the halls of the U.S. Congress and Egyptian parliament. As the April 23 edition of Newsweek put it, "A brief encounter at a Cairo cocktail party could signal a shift in Bush administration policy toward the Muslim Brotherhood, a worldwide Islamic movement that the United States has shunned because of its alleged ties to terrorism."

The potential scandal began in March when the press first detailed an American congressional delegation's visit to Egypt, led by the House majority leader, Rep. Steny Hoyer, a Democrat of Maryland, who had multiple meetings with members of the Muslim Brotherhood.

The meetings took place at the Egyptian parliament and at the home of an American ambassador, Francis Ricciardone. A spokesman for the Brotherhood, Hamdi Hassan, told the Associated Press that developments in the Middle East and his organization's ideology were discussed in private.

In an interview on the Muslim Brotherhood's Web site, Ikhwanweb.com, on April 12, Muhammed Sad Al-Katani, the chairman of the Brotherhood's bloc in parliament and one of those who met with the American delegation, added that it comprised seven Democrats and four Republicans.

By all accounts and to his credit, Mr. Hoyer argued during the meetings with the Brotherhood members over issues related to terrorism and Hamas.

Yet the Egyptian opposition daily Al-Masryoon reported April 12 that as a follow-up to the initial meetings in Egypt, "high-level diplomatic sources" have said the Muslim Brotherhood general guide, Muhammad Madi Akef, several members of his office, and Muslim Brotherhood MPs had been invited by an unidentified congressman to visit America next month and to speak to Congress. The Muslim Brotherhood has recently denied the invitation.

Al-Masryoon then reported April 16 that Mr. Ricciardone left Egypt secretly last Thursday, after being summoned by the State Department to discuss the ramifications of the recent meeting between Mr. Hoyer and Mr. Katani. The sources added that Egypt's Foreign Ministry had informed Foggy Bottom of its objections to this meeting because it gives legitimacy to a movement that is outlawed in Egypt.

A lack of knowledge about the Muslim Brotherhood is evident on the part of U.S. officials who are now cozying up to the organization. The Brotherhood's leadership is vehemently anti-American and has gone on the record calling for jihad against America.

One notable example includes Brotherhood lawmaker Ragab Hilal Hamida's speech during a parliamentary session between January 28 and February 3, 2006, which focused on the Inter-Arab Agreement on Combating Terrorism. The Arab press reported on his statements saying the Koran encourages terrorism and that he supports the activities of Osama bin Laden, Ayman Al-Zawahri, and Abu Musab Al-Zarqawi. The report also stated that another Muslim Brotherhood lawmaker, Sheikh Ahmad Askar, agreed with Mr. Hamida's statements.

Another example illustrating the problem with the Muslim Brotherhood can be found in the ideology of its current leader. When Mr. Akef took over in 2004, he gave a series of interviews to the Egyptian press stating — among other things — that America, which he referred to as "Satan," would soon collapse. Regarding the attacks of September 11, 2001, he claimed there was no real proof that Al-Qaeda was behind them and stated that in the future, Islam would take over Europe and America.

An unidentified senior official in the American government was quoted by Newsweek as saying the invitation of Brotherhood members was "cleared" by the State Department and represented the highest-level contact with the organization since September 11.

As some members of the American government pursue talks with the Muslim Brotherhood, it must be understood that this organization is no friend of America. Just two weeks ago, its leader spoke at a Cairo-based conference called "The Struggle Against the U.S. and Zionist Occupation." He stated that "the devil Bush" and his allies were really behind "sowing terror worldwide."

Hamas's Western Cheerleaders

April 25, 2007

> *"Israel will exist, and will continue to exist, until Islam abolishes it, as it abolished that which was before it.' [From the words of] the martyr, Imam Hassan al-Banna, Allah's mercy be upon him."*
>
> — OPENING PARAGRAPH OF THE HAMAS COVENANT

Since the Palestinian Arab unity government was established in March, some in the West have made a concentrated effort to speak in defense of "a new, improved Hamas."

Diplomats from European coun tries, led by Norway, have begun to meet regularly with representatives of the terrorist organization. Switzerland and China also have announced that they will now deal with Hamas.

Even Western writers have be gun to discuss this new Hamas which is "ready for peace." Robert Novak's "Olive Branch From Hamas," an April 16 Washington Post column, is one example.

Hamas's cheerleaders in the West, including Mr. Novak, have been quick to accept everything the organization says in English about its peaceful intentions while ignoring statements in Arabic about its plans to "slaughter the Jews."

A case in point is a Hamas spokesman and frequent inciter to suicide attacks during the second intifada, Ismail Radhwan. After a long hiatus, Mr. Radhwan returned to official Palestinian Authority TV on March 30 to give the Friday sermon, preaching that Islamic tradition dictates that Muslims must fight and kill the Jews.

Mr. Radhwan regularly led Friday sermons when suicide attacks in Israel were at an all-time high, and he discussed the virtues of "martyrdom operations" and what happens to a "shahid," or suicide bomber when he detonates himself.

Even when "a martyr's organs are being chopped off and he turns into torn organs that spread all over, in order to meet Allah, Muhammad and his friends, it would not be [considered] a loss," Mr. Radhwan said as he began his sermon at the Sheikh Ijlin Mosque in Gaza on August 17, 2001 — just a week after the Sbarro pizzeria bombing in Jerusalem that killed 15 people, seven of them children.

Last week, as the voices of Hamas's Western cheerleaders grew louder, the acting speaker of the Palestinian Legislative Council and a Hamas member, Ahmad

Bahr, visited Sudan and was given the honor of leading the Friday sermon on live Sudanese television.

Note that his official duty is to speak for the Palestinian Arab government. In the sermon, Mr. Bahr said, "America and Israel will be annihilated, Allah willing."

Those capable of making such a statement clearly are not capable of carrying an olive branch.

War Crime Warrants Revive 'Darfur Conspiracies'

May 3, 2007

The International Criminal Court in Amsterdam yesterday issued arrest warrants for a Sudanese minister and a militia leader suspected of war crimes in Darfur.

The warrants against the Sudanese humanitarian affairs minister, Ahmed Haroun, and a Janjaweed leader, Ali Muhammad Ali Abd al-Rahman (also known as Ali Kushayb), detail 51 counts of murder, rape, torture, and forced displacement of villagers, among other crimes.

Yet even as the genocide in Sudan officially becomes a crime against humanity, many leaders and writers in the Arab world continue to deny the facts about what is happening and foment conspiracy theories, some of them anti-Semitic, to explain international outrage over the situation.

Members of the Sudanese government have taken the lead in simply contradicting news about Darfur. When the International Criminal Court prosecutors first filed charges against Messrs. Haroun and Abd al-Rahman earlier this year, President Bashir of Sudan called allegations of crimes against humanity in Darfur a "fabrication" in a March 19 interview on NBC.

Other apologists for Sudan recognize that the grimness of the situation there is too well documented to deny outright. On Monday, the secretary-general of the Arab League, Amr Moussa, responded to an April 25 House resolution calling on the Arab League to recognize that the conflict in Darfur is a genocide by calling the resolution "far from reality" and expressing "great astonishment" that the U.S. Congress would demand "that the Arab League declare that the human rights violations taking place in Darfur are genocide, while no international or regional organization is using this definition."

Similarly, an influential former editor of the London Arabic daily Al-Hayat, Jihad Al-Khazen, wrote on April 13 that, while killing in the region was occurring, "the Israel lobby" was to blame for making the situation worse than it actually is.

In a piece headlined "Since the Victims Are Arabs and Muslims," Mr. Khazen wrote: "In New York, Darfur is the most important issue in the world, or at least this is what the resident or visitor sees and hears. From subway tunnels to the streets, there are thousands of posters talking about 'genocide' and 400,000 people killed."

But "the lobby to save Darfur," he argued, is inflating the casualty count. "Darfur is a terrible humanitarian disaster that should not be played down. I am not doing that myself. However, the United Nations itself said that 200,000 were killed and that what had been committed there were war crimes, not genocide," he wrote.

Why would the "Darfur lobby" do this? Because, according to Mr. Khazen, "the lobby to save Darfur is just the Israel lobby renamed. The goal is to divert attention from Israel's crimes, or the catastrophe of the war in Iraq."

An English-language apologist for the Khartoum government has echoed another version of this theory that is frequently expressed in the Arab press. Sara Flounders, who has worked on Sudanese issues for a former attorney general under President Johnson and attorney for Saddam Hussein, Ramsey Clark, wrote in Sudan Vision: "This Darfur campaign accomplishes several goals of U.S. imperialist policy. It further demonizes Arab and Muslim people. It diverts attention from the human rights catastrophe caused by the brutal U.S. war and occupation of Iraq. … It is also an attempt to deflect attention from the U.S. financing and support of Israel's war on the Palestinian people. Most important, it opens a new front in the determination of U.S. corporate power to control the entire region."

As this column first documented in "The Darfur Conspiracy" in fall 2005, the Arab press has been rampant with conspiracies surrounding Darfur ever since news first emerged about the atrocities in Sudan. As the esteemed head of Al-Arabiya TV, Abd al-Rahman al-Rashed, wrote in the London Arabic daily Asharq Al-Awsat on June 24, 2004, the writing concocting such diversionary tactics are "intellectual accomplices in the crime."

What such Darfur apologists are condoning, according a report from the Watch List on Children and Armed Conflict released on April 20, are "unspeakable acts of violence and abuse," including the kidnapping, rape, and torture of Sudanese children.

One of the authors of the report adds, "Children in Sudan continue to endure some of the most inhumane treatment found anywhere in the world."

So, as many in the West clamor to take action to stop the genocide in Darfur, the anti-American and anti-Semitic politicians and writers in the Arab world have embraced "Darfur denial" as a means to further their agendas. Unfortunately, this phenomenon is only growing.

Islamist-Left Alliance A Growing Force

May 9, 2007

"Where else can you sit down in a single evening and listen to senior people from Hamas, Hezbollah, the Muslim Brotherhood, people from the revolutionary left and the antiwar movement from around the globe?"

— BRITISH TROTSKYITE JOHN REES AT THE CAIRO ANTI-WAR CONFERENCE, APRIL 2007

Over the past year, multiple international conferences have featured leaders of the anti-global left and Islamist groups working together. Go to any anti-war or anti-globalization demonstration in the West and chances are you will see the flags of Hezbollah and Hamas waved by people wearing Che Guevara T-shirts. And at some of these meetings, members of such radical Islamist groups as the Muslim Brotherhood, Hamas, and Hezbollah have enjoyed starring roles.

The roster of Islamist-left alliances quietly grows every day: Massachusetts Institute of Technology linguistics professor Noam Chomsky praises Hamas and denounces America on Hezbollah's Al-Manar television. London Mayor Ken Livingstone invites a leading Islamist, Sheikh Yosef Al-Qaradawi, who is known for supporting suicide attacks, to visit his city. Iranian President Ahmadinejad calls for a world without America even as he plays host to a Tehran peace conference attended by American Mennonites, Quakers, Episcopalians, Methodists, and leaders of the National Council of Churches.

The key forum at this year's annual Cairo Anti-War Conference was titled "Bridge-building Between the Left and Islam," and focused on practical ways to increase cooperation. The aim of the conference sessions were described in one piece of literature as tackling "the challenges and prospects facing the international anti-war and pro-intifada movements" and planning "strategy and tactics for bridging the gap and uniting Islamist and leftist ranks in the face of U.S. imperialism and Zionism."

The Arabic press has lauded this phenomenon. An article in Egypt's Al-Ahram Weekly praises what it refers to as "Arab activists taking the lead in the growing anti-war movement worldwide."

Some Islamist issues on which the anti-global left appears to have found common ground are countering the international boycott of Hamas, calling for the boycott of Israel, supporting Iran against the threat of a U.S. attack (as well as supporting its "right" to develop nuclear weapons), supporting avowedly anti-American countries such as Cuba and Venezuela, and, above all, opposing the war in Iraq.

Many leftist blogs and Web sites monitored by MEMRI have been increasingly open to working with Islamist groups on the goals they both share — most notably the desire for America to leave Iraq. British Trotskyite John Rees, a regular at the Cairo conference since it started in 2002, told Al-Ahram in April that at this year's gathering he had discussed in particular the "ongoing dynamic between the anti-war movement and Egypt's Muslim Brotherhood." Mr. Rees also believes that the Islamist-Left coalition is gaining strength, while what is known as the "coalition of the willing" has faltered.

In addition, MEMRI researchers monitoring jihadi Web sites have recently found Islamists trying to influence American anti-war efforts. On the Islamist Al-Mohajroon Web site, someone with the username Al-Wathiq Billah instructs readers on how to infiltrate popular American Internet forums to distribute jihadist films and spread disinformation about the war.

"There is no doubt, my brothers, that raiding American forums is among the most important means of obtaining victory in the fierce media war ... and of influencing the views of the weak-minded American who pays his taxes so they will go to the infidel American army. This American is an idiot and does not know where Iraq is ... Every electronic mujahid" must engage in this raiding, Mr. Billah writes.

Mr. Billah advises his jihadist readers to "register yourself using a purely American name" and to "invent stories about American soldiers you have personally known (as classmates... or members in a club who played baseball and tennis with you) who were drafted to Iraq and then committed suicide while in service by hanging or shooting themselves." The writing should, he says, provoke "frustration and anger towards their government, which will ... render them hostile to Bush ... and his Republican Party, and make them feel they must vote to bring the troops back from Iraq as soon as possible."

Such an emerging alliance can only be expected to play a negative role in the ongoing war on terror.

Mullah Gets His Wish For Martyrdom

May 16, 2007

> *"The America of today is not the America of the past. ... [It] has sunk so low that even the widows now want to participate in the war against it, and wish to defeat the U.S. ... We will bring shameful defeat upon the Jews and the Christians. The cries you are now hearing from them once a day, you will hear them 20 times a day."*
>
> — MULLAH DADULLAH, AL-JAZEERA TV, MARCH 2

The demise of the Taliban's top military commander, Mullah Dadullah, is certainly a victory in the war on terror. The one-legged Dadullah was by all accounts a vicious tyrant personally responsible for beheadings and other outrages who had rightfully earned his nicknames: the "Wild Beast" of the Taliban and the "Butcher of Kandahar."

In Dadullah's interviews, he was often quoted explaining his desire to kill "Jews and Christian forces." Not surprisingly, jihadists are celebrating his death as that of a martyr who will soon receive his 72 black-eyed virgins. On one Islamist Web site, a poem compares the mullah to a companion of Islam's Prophet Muhammad during the Battle of Uhud in the year 625.

Soon after Dadullah's martyrdom was announced, MEMRI's team of Islamist Web site monitors found online chatter that speculated that his death might be connected to his love for being in front of TV cameras. His frequent interviews on Al-Jazeera had been criticized by many in the Taliban, and on January 19, Islamist Web sites posted a communiqué purported to be from the Taliban in which a spokesman identified as "Dr. Hanift" claimed, "The Taliban organization may have issued an order to kill Mullah Dadullah."

Rahimullah Yusufzai, an editor of a Peshawar, Pakistan-based newspaper, the News, yesterday was quoted stating that Taliban fighters had been unhappy with Dadullah "because of his beheadings, kidnappings, and boastful videos."

Dadullah was a lover of martyrdom and would often give sermons espousing suicide bombing for Islam. An Islamist Web site that eulogizes Dadullah claims he frequently called for himself to come back to life again and again in order to be able to be martyred 100 times.

The mullah was even known to distribute "tickets to heaven," vouchers for volunteer suicide bombers. According to accounts, he tried to create innovative ways to help young men "to accept martyrdom for the sake of jihad." In his last interview, given a week ago, Mr. Dadullah warned that Americans and British citizens "will be executing attacks in Britain and the U.S."

In a May 31, 2006, interview on Al-Jazeera, Dadullah boasted, "Praise be to Allah, who gave us this great power of self-sacrifice, among Arabs and non-Arabs. Our preparations for the war in the field of self-sacrifice activities are not enough by themselves. … The weapon of devotion is still the most important. In addition, we must remember our martyrs and their blood."

A video from As-Sahab, the production house for Al-Qaeda, that was posted at tajded.net.tc on December 28, 2006, is centered on Dadullah and opens with chants of "Oh, Muslim, arise! There is a black-eyed virgin in Paradise. … Regards to the Taliban, oh, blessings of the All Merciful." The video conflates its cause with "the Sunnah after the Koran," which is the "trodden path" of Muhammad's teachings. Dadullah continues, referring to Osama bin Laden and President Bush by name: "You are sufficient, oh, bin Laden. … The rule of jihad today does not require much thought, oh, Muslim. … Oh, Bush, you are facing men for whom, oh, Crusader, the sound of the bazooka is like music."

In the As-Sahab video, Dadullah is shown greeting a group of suicide bombers on their way to attacks. "May Allah bless you," he says to the prospective killers as they are sent on their way.

Now that Dadullah has joined Abu Musab al-Zarqawi, one hopes other terrorists such as Mr. bin Laden will follow.

A Muslim Friend of America

May 23, 2007

> *"I am one of the supporters of the alliance with America. ... I support the strengthening of the alliance with America —military, cultural, and economic alliance — because I believe that the Gulf has no future without this alliance. All the leaders of the ...Gulf Cooperation Council have declared that their alliance with America is strategic."*
>
> — ABD AL-HAMID AL-ANSARI ON AL-ARABIYA TV, MAY 1

A leading intellectual from the Arab and Muslim world, Abd Al-Hamid Al-Ansari has a long list of impressive credentials. Not only was he dean of the University of Qatar, he also headed its Islamic law department. His writings frequently appear in leading Arabic newspapers, such as the London daily Asharq Al-Awsat.

Mr. Ansari can often be found debating Islamists on the main Arab satellite channels. Fighting "the Arab culture of hatred" is at the heart of his work, most notably his support for revamping Arab school curriculums and purging them of hate.

His voice was one of the first from the Arab world denouncing the attacks of September 11, 2001. In an article in the pan-Arab daily Al-Hayat on December 4, 2001, in what could be considered a laying of the groundwork for the emergence of the Arab reform movement, he asked whether the Arab "satellite channels have the right to broadcast terrorist operations and incitement on the pretext of the 'principle of freedom for all,' disseminating hatred in viewers who then undertake harmful acts of stupidity. We must examine our curriculum and evaluate our educational methods. We must re-examine our education and our media. This will be the right beginning for the fight against the culture of terrorism."

Following the release of the September 11 commission report in July 2004, Mr. Ansari wrote in another Al-Hayat article that "The 9/11 Commission depicts the full and accurate picture of the attacks [and] their planning. ... No doubt the Americans are going to draw practical lessons from the report in order to prevent the recurrence of their failings. But what about us?"

Mr. Ansari went on to criticize Arab intellectuals and leaders for blaming the attacks on Israel, Jews, and America, and called on them to "apologize for their words, mistakes, and for misleading other people now that all the facts have been clarified. ... Do we have the courage to criticize ourselves, to admit to our faults, and to apologize?"

He asked Arabs also to also draw their own practical lessons from the report. "Why won't we take the opportunity of the appearance of the 9/11 Commission Report to ponder why destructive violence and a culture of destruction have taken root in our society? Why won't we take this opportunity to reconsider our educational system, our curricula, including the religious, media, and cultural discourse that causes our youth to live in constant tension with the world?" he wrote.

Around the time America began its operations in Afghanistan, Mr. Ansari gave an extensive interview to the Qatari daily Al-Raya in which he supported America's right to use force against those providing a safe haven for the planners of the September 11 attacks: "It is unfair to name the American response 'terrorism,' because by so doing we are confusing the concepts of terrorism and self-defense. ... What happened in America is terrorism; the American response is a response to that aggression."

Of the Taliban, he has said: "Any country or group protecting and defending terrorists must be fought, and the world must be saved from their evil. ... The time has come to call them to account, and punish them. ... We must have the courage to admit that what happened in Afghanistan was the liberation of our Muslim brothers."

When Saddam Hussein was deposed, Mr. Ansari wrote an optimistic message about the future of the Middle East in the Saudi daily Arab News on April 21, 2003: "The Iraqi people brought down the statue of the tyrant who had oppressed them for 35 years. Millions saw the wretched end to one of the most loathsome and bloody regimes in modern history. ... There are other nations who have suffered defeat and have risen from its ashes within a single generation."

Earlier this month, Mr. Ansari appeared on Al-Arabiya TV and was asked if he still believed that the American presence in Iraq is legitimate. The former legal educator answered, "It is legitimate and legal" and said it is not in the interests of Iraqis for the Americans to leave now. He added, "Iraq needs the American presence for its stability and defense."

Mr. Ansari is just one of many Arab reformists whom the West should support by all means. To read more about him and others like him, visit www.memri.org/reform.

The Islamist Threat Arrives in Switzerland

May 30, 2007

> *"Salman bin Fahd Al-Odah is a preacher of global influence and is one of the senior figures of the fundamentalist Islamic Wahhabi movement in Saudi Arabia as well as a close associate of Osama bin Laden. He was imprisoned in Saudi Arabia for his extremist ideologies from 1994 until 1999. Even after his imprisonment, he adopts the call to armed struggle against the infidel Western countries in his writings."*
>
> — A REPORT FILED BY THE MINISTRY OF JUSTICE IN SWITZERLAND, MAY 2007

In another sign of the spread of Islam within Europe, last month, Swiss Muslims announced their plan to open "Europe's biggest Islamic center" in the capital city of Bern. The center is estimated to cover 84 acres and to cost as much as $66 million to build.

Switzerland seems to be an unlikely locus for a battle over jihadist Islamism, but according to reports, its Muslim citizens, who make up about 5% of the total population, increasingly look to radical Middle East clerics for spiritual guidance. The country is also home to a controversial professor, Tariq Ramadan, whose visa to come to America was revoked by the State Department, and reports indicate there has been a rising tension between Muslims and non-Muslims there. In what could be considered pouring fuel on a fire, two weeks ago, the League of Swiss Muslims invited Sheikh Salman bin Fahd Al-Odah of Saudi Arabia to participate in their annual conference.

Hailing from a wealthy Saudi family, Mr. Odah is a noted Islamic scholar who has a wide following among Islamists throughout the world. He was enraged when American troops were based in Saudi Arabia during the first Gulf War and has gone on record supporting jihad against American troops in Iraq.

His stance on the American presence in Saudi Arabia earned him five years in jail there for speaking out against the government — a sentence condemned in Osama bin Laden's infamous 1996 fatwa against the West. According to reports, copies of Mr. Odah's sermons have been found in an abandoned Afghanistan home belonging to Mr. bin Laden.

When Al-Qaeda in Saudi Arabia released the first issue of its online magazine, "The Voice of Jihad," in 2003, it included a biography of Sheikh Yousef Al-Ayyiri, a Saudi Al-Qaeda leader who served as Mr. bin Laden's personal bodyguard in Afghanistan and later in Sudan. Upon Mr. Ayyir's return to Saudi Arabia, that issue of "The Voice of Jihad" quotes Mr. Al- Odah as telling Mr. Ayyiri, "I am proud to be one of the soldiers of Abu Abdallah," Mr. bin Laden's nom de jihad.

A few weeks ago, as Mr. Odah was about to leave for Switzerland, Swiss authorities informed him he had been barred from entering Switzerland by a ruling of the Federal Police Department of the Swiss Ministry of Justice. Accusing "extremist Zionist forces" of being behind the ban, Mr. Odah said he is considering a lawsuit and says that American and German lawyers have agreed to take the case.

According to a London Arabic daily, Al-Asharq Al-Awsat, Mr. Odah "was subjected to a similar incident in Spain, where Spanish authorities accused him of financing one of the criminals in the Madrid train bombings." Publications such as El Mundo in Spain and Corriere della Sera in Italy have alleged that Madrid bomber Rabei Osman Ahmed was a follower of Mr. Odah.

In the past year, Mr. Odah has made a number of negative headlines in the West. He attacked Pope Benedict's statements about Islam. The Baltimore Sun reported on January 8 that Mr. Odah was "a regular visitor to Sudan and a strong supporter of the Janjaweed's campaign against the people of Darfur."

The Baltimore Sun article also accused him of "funding the Islamic Courts Union" in Somalia, a political group that has tried to turn Somalia into a Wahhabist-style state, and reported that Mr. Odah had "called upon Saudis to go to Somalia to aid the jihad" there during a December 27 interview on Al-Jazeera television.

It is however a troubling sign that such Muslim leaders have become religious and spiritual guides to mainstream Muslims in the West.

Switzerland's banning Mr. Odah follows decisions in other Western countries to close their borders to influential Muslim religious figures who often espouse anti-Western and pro-jihad sentiment. Notable examples include America's denial of entry to a leading Egyptian Islamist sheikh, Yusef Al-Qaradawi, a spiritual leader of the Muslim Brotherhood, and Canada's barring the imam of Islam's holiest mosque in Mecca, Abdul Rahman Al-Sudais, for his anti-Semitic and anti-Christian statements and calls for jihad. This practice should continue and be adopted by all Western countries.

Topical Index

ISLAMISM IN THE WAR ON TERROR

CONSPIRACY THEORIES

REFORM IN THE MIDDLE EAST

JIHAD/ISLAMISTS IN THE WEST

JIHAD TV IN THE WEST

CAIR AND JIHAD

ANTISEMITISM AND HOLOCAUST DENIAL IN THE ARAB AND MUSLIM WORLD

QATAR'S AL-JAZEERA TV

HEZBOLLAH AND ITS PROPAGANDA EFFORTS

HEZBOLLAH VS. ISRAEL

SAUDI TV

THE SAUDI TERROR CONFERENCE

SAUDI ANTI-SEMITIC AND ANTI-WEST INCITEMENT

THE IRAN THREAT

PALESTINIAN MARTYDOM, INTIFADA, AND HAMAS

EGYPT AND JIHAD

DARFUR CONSPIRACY THEORIES

ARABS ON U.S. POLITICS AND POLITICIANS

ARAB ANTI-CHRISTIAN RHETORIC

www.ingramcontent.com/pod-product-compliance
Lightning Source LLC
LaVergne TN
LVHW021743071126
851857LV00039B/599